Analog Circuits and Signal Processing

Series Editors

Mohammed Ismail
Mohamad Sawan

For further volumes:
http://www.springer.com/series/7381

Analog Circuits and Signal Processing

Francisco Aznar · Santiago Celma
Belén Calvo

CMOS Receiver Front-ends for Gigabit Short-Range Optical Communications

 Springer

Francisco Aznar
Faculty of Sciences
University of Zaragoza
Pedro Cerbuna 12
50009 Zaragoza
Spain

Belén Calvo
Faculty of Sciences
University of Zaragoza
Pedro Cerbuna 12
50009 Zaragoza
Spain

Santiago Celma
Faculty of Sciences
University of Zaragoza
Pedro Cerbuna 12
50009 Zaragoza
Spain

ISBN 978-1-4899-8669-6   ISBN 978-1-4614-3464-1 (eBook)
DOI 10.1007/978-1-4614-3464-1
Springer New York Heidelberg Dordrecht London

Printed on acid-free paper

Springer is part of Springer Science+Business Media (www.springer.com)

Preface

Nowadays, long-haul broadband data transmission is dominated by optical communications due to its superior performances in terms of speed and reach compared to its electrical counterparts. Charles Kao was awarded with the Nobel Prize in 2009 owing to demonstrate the viability of such systems before its development and widespread use. In recent years, the transition from electrical to optical transmission for short-reach applications has begun. The market is demanding such transmissions for local area networks, broadband internet connection (fiber-to-the-home), multimedia systems inside cars and homes and, in the future, interconnections between electronic devices on the same board and, even, between cores of the same processor. The main reason is to overcome the speed limitation derived from electrical transmission, although there also exist other significant advantages, such as avoiding electromagnetic interference, galvanic separation and improvements in security.

The low cost of the transmission system is a must, as the number of users is very limited, or even, single-user. In this new scenario, plastic optical fiber (POF) instead of glass optical fiber and CMOS technology are the best suitable candidates to implement the transmission link due to their lower fabrication cost. The overall key idea is to implement the analog front-end in the same low-cost technology as the subsequent digital circuitry, leading to integrate the entire receiver system in only one CMOS chip. Moreover, CMOS design has been further encouraged with the advent of compatible light emitters and the development of silicon photodetectors, in order to achieve a system-on-chip architecture, that is, a transceiver fully integrated in CMOS technology.

Therefore, the main objective of this book is to develop optical receiver solutions integrated in standard CMOS technology, attaining high-speed short-range transmission under a cost-effective constraint. Thus, the focus is on reliable full CMOS receiver solutions targeting gigabit transmission along a low-cost standardized plastic optical fiber up to 50 m length. The aim is offering a detailed study of the design constraints and challenges arisen under these conditions—both at basic cells and system level—, their main characteristics and optical and electrical performances.

The starting point is a complete review and analysis of optical communications, covering from an historical introduction to the current state of the art and detailing emerging applications, issues addressed in Chap. 1. Next, the fundamentals of optical signal transmission are explained, i.e., characteristics of data signal, introduction to optical fibers, review of the complete transmission system and the key parameters of an optical receiver from receiver's point of view. Thus, the most relevant aspects of the entire optical link are covered in Chap. 2, analyzing the characteristics and limitations affecting receiver of every element: output power and extinction ratio of laser, loss and bandwidth of optical fiber and responsivity and capacitance of photodetector. In Chaps. 3 and 4, the analysis, design, implementation and test of the main building blocks of the optical receiver—transimpedance amplifier and post-amplifier—under low voltage restrictions are exhaustively described, respectively. Some unique blocks, techniques and architectures proposed by the authors are offered, worth mentioning the technique to increase the dynamic range of the transimpedance amplifier and the AGC post-amplifier approach. A 0.18 μm standard CMOS technology is mainly used to achieve a cost-effective solution. In addition, 90 nm CMOS technology is explored to improve transimpedance amplifier's performances, the most challenging structure as it directly impacts receiver's performances. To close, in Chap. 5, a gigabit transmission over a standardized cost-effective plastic optical fiber up to 50 m reach has been attained for a complete receiver formed by the two proposed building blocks and a mandatory adaptive equalizer to compensate the limited and length-dependent bandwidth of such a fiber. Measured performances leads to an eye safety operation, required by an inexpensive do-it-yourself installation, up to almost 30 m reach. These contributions prove that this approach is a more than competitive alternative in the upcoming market-demanded short-range applications where economic viability is a must, such as fiber-to-the-home, local area networks and multimedia systems in cars or homes, evidencing that the class of solutions advanced in this book will take an important role in the near future.

We would like to thank the Barcelona Microelectronics Institute attached to National Microelectronics Center for measurement support. We must remark the collaboration of Fortia Vila, responsible of the on-wafer probe station. We gratefully thank the Institute of Electrodynamics, Microwave and Circuit Engineering attached to Vienna University of Technology for the cooperation with this research. In particular, two fruitful research stays were carried out supervised by the head of the institute, Prof. Horst Zimmermann. F. Aznar thanks Prof. Zimmermann and all colleagues from EMCE for their help and support during research stays, especially Robert Swoboda, Wolfgang Gaberl, Kerstin Schneider-Hornstein, Franz Schlögl and Bernhard Goll.

Finally, we would like to thank to the institutions which have financially supported this work: Ministry of Science and Innovation through projects in collaboration with the European Regional Development Fund (TEC2008-05455 and

TEC2011-23211) and the PhD fellowship (AP2006-01434), and the saving bank *Caja de Ahorros de la Inmaculada* which have supported the research stays through the CAI-Europe Program (IT5/08, IT21/09).

Zaragoza, Spain, February 2012

Francisco Aznar
Santiago Celma
Belén Calvo

Contents

Symbols

Constant	Description	Value	Unit
K	Boltzmann's constant	1.38065×10^{-23}	J/K
h	Planck's constant	6.626×10^{-34}	J s
q	Electron charge	1.6022×10^{-19}	C
ε_0	Vacuum dielectric constant	8.8542×10^{-12}	F/m
μ_0	Vacuum magnetic permeability	$4\pi \times 10^{-7}$	H/m
c	Vacuum light speed	299,782,465	m/s

Symbol	Description
BW	Bandwidth
BW_N	Noise bandwidth
C_F	Feedback or filtering capacitance
C_{gs}	MOS gate-source capacitance
C_I	Input capacitance
C_L	Load capacitance
CLK	Clock signal
C_O	Output capacitance
C_{OX}	Oxide capacitance per unit of area
C_{PD}	Photodiode capacitance
DL	Decision level
EN	Enable
F	Excess noise factor
f_{CLK}	Clock frequency
f_{LF}	Low-frequency cut-off
f_T	Transition frequency
G or A	Gain
g_m	MOS transconductance
G_m	Cell transconductance
GND	Ground

$H(s)$	Transfer function
I_B	Bias current
I_{DS}	Drain-source current
I_H	Current for high state
I_{IN}	Input current
I_L	Current for low state
I_N	Noise current
I_{OFF}	Leakage current under specific conditions
I_{ON}	I_{DS} current under specific conditions
j	Complex
L	Transistor length
n	Refraction index
P_{AV}	Average power
P_H	Power for high state
P_{IN}	Input power
P_L	Power for low state
P_{LPF}	Power of low-pass filtered signal
P_{OV}	Overload power
P_{TOT}	Power of entire signal
Q	Q factor
R	Responsivity
R_b	Bit rate
R_F	Shunt-feedback or floating load resistance
R_L	Load resistance
S	Sensitivity
s_a	Dominant pole
S_{xy}	S parameters (x and y = 1 or 2)
T	Temperature
T_b	Pulse width
T_{OX}	Gate-oxide thickness of a MOS transistor
T_R	Transresistance
v	Velocity
V_B	Bias voltage
V_C	Control voltage
V_{CC}	Supply voltage
V_{CM}	Common-mode voltage
V_E	Error signal
V_{INI}	Initializing signal
V_L	Low-pass filtered voltage signal
V_{PD}	Peak detected voltage
V_{REF}	Reference voltage

V_{SW}	Voltage swing
V_T	Entire voltage signal
V_{TH}	Transistor threshold voltage
W	Transistor width
ε	Dielectric constant
ζ	Damping ratio
η	Quantum efficiency
θ	Angle
θ_L	Limit angle
λ	Wavelength
μ	Magnetic permeability
μ_N	Electron mobility
μ_P	Hole mobility
σ	Standard deviation
τ	Time constant
φ	Probability density
χ	Efficiency slope
Ψ	Wave function
ω	Angular frequency
ω_0	Characteristic frequency

Acronyms

Acronym	Significance
AC	Alternate Current
AGC	Automatic Gain Control
ASIC	Application-Specific Integrated Circuit
ASK	Amplitude-Shift Keying
BER	Bit Error Rate
BERT	Bit Error Rate Tester
BIPM	*Bureau International des Poidset Mesures*
CDR	Clock and Data Recovery circuit
CG	Common-Gate
CMOS	Complementary Metal-Oxide-Semiconductor
DC	Direct Current
DCA	Digital Communications Analyzer
DFB	Distributed FeedBack
DMUX	DeMUltipleXer
DR	Dynamic Range
DSP	Digital Signal Processor
DUT	Device Under Test
DWDM	Dense Wavelength-Division Multiplexing
EMI	Electro-Magnetic Interference
ER	Extinction Ratio
FOM	Figure Of Merit
FET	Field Effect Transistor
FF	Flip-Flop
FP	Fabry-Perot
FSG	Fluorine doped Silicate Glass
FTTH	Fiber-To-The-Home
GBW	Gain-Bandwidth Product

GI	Graded-Index
GOF	Glass Optical Fiber
GSG	Ground-Signal-Ground
HPF	High-Pass Filter
HVT	High Threshold Voltage
IC	Integrated Circuit
IDC	International Data Corporation
IEEE	Institute of Electrical and Electronics Engineers
ISI	Inter-Symbol Interference
LA	Limiting Amplifier
LAN	Local Area Network
LASER	Light Amplification by Stimulated Emission of Radiation
LED	Light-Emitting Diode
LFSR	Linear Feedback Shift Register
LL	Low Leakage
LPF	Low-Pass Filter
LVT	Low Threshold Voltage
MAN	Metropolitan Area Network
MIM	Metal-Insulator-Metal
MLS	Maximal Length Sequence
MM	Multi-Mode Fiber or Mixed Mode
MOSFET	Metal Oxide Semiconductor Field Effect Transistor
MOST	Media Oriented System Transport
MUX	MUltipleXer
NMOS	Metal-Oxide-Semiconductor N
NRZ	Non Return to Zero
PA	Post-Amplifier
PAM	Pulse Amplitude Modulation
PCB	Printed Circuit Board
PD	PhotoDiode or Peak Detector
PGA	Programmable Gain Amplifier
PLL	Phased-Locked Loop
PMMA	Poly-Methyl MethAcrylate
PMOS	Metal-Oxide-Semiconductor P
POF	Plastic Optical Fiber
PP	Power Penalty
PRBS	PseudoRandom Bit Sequence
PSK	Phase-Shift Keying
PVT	Process-Voltage-Temperature
QFN	Quad-Flat No-lead package
RCLED	Resonant Cavity Light Emitting Diode

RF	RadioFrequency
RMS	Root-Mean-Squared
RVT	Regular Threshold Voltage
RZ	Return to Zero
SI	Step-Index
SM	Single-Mode Fiber
SMA	SubMiniature – Version A
SNR	Signal to Noise Ratio
SOLT	Short Open Load Through
SP	Standard Performance
SR	Shift Register
TAT	TransAtlantic Telecommunications cable
TIA	TransImpedance Amplifier
UMC	United Microelectronics Corporation
USB	Universal Serial Bus
VCO	Voltage-Controlled Oscillator
VCSEL	Vertical-Cavity Surface-Emitting Laser
VGA	Variable Gain Amplifier
VNA	Vector Network Analyzer
WDM	Wavelength-Division Multiplexing

Chapter 1
Introduction

Physics has been characterized by a continuous progress. In this progress, some remarkable leaps can be identified over the ages that have led to the rapid development of particular fields of knowledge. The first great physicist who comes to our mind is Isaac Newton who is considered the most important scientist of all times. His work, *Philosophiæ Naturalis Principia Mathematica* (Mathematical Principles of Natural Philosophy), published in 1687, includes two main scientific contributions: universal gravitation and the laws of motion. Both establish the basis for classical mechanics that governed the scientific view of the physical universe for the next three centuries until relativity was confirmed.

Based on Newton's mechanics, Gottfried Leibniz postulated the conservation of a quantity connected with motion denominated *vis viva*. In 1783, Antoine Lavoisier and Pierre-Simon Laplace reviewed the concept of *vis viva* and the caloric theory, suspecting that there was a relationship. Eventually, the conservation of energy was described and broadly assumed. Thomas Young introduced the well-known concept of energy in 1807. The generalization of the conservation of energy is one of the basic principles of thermodynamics and it is still valid.

In the nineteenth century, three research fields of physics were under study: electric and magnetic fields, and light. James Clerk Maxwell expounded the unified field theory based on the laws introduced by Carl Friedrich Gauss, André-Marie Ampère, and Michael Faraday. Maxwell's equation appears for the first time in the article "A Dynamical Theory of the Electromagnetic Field", published in 1864. The unification of the electric and magnetic fields predicts an electromagnetic wave traveling at the speed of light thereby confirming the wave behavior of light as defended by Christiaan Huygens.

The last leap in physics was achieved by the most important physicist of the twentieth century and the most famous scientist ever, Albert Einstein. In 1905, he postulated the special theory of relativity in one of the *Annus Mirabilis papers* entitled *Zur Elektrodynamik bewegter Körper* (On the Electrodynamics of Moving Bodies), which is considered the starting point of "modern physics". Then, in

F. Aznar et al., *CMOS Receiver Front-ends for Gigabit Short-Range Optical Communications*, Analog Circuits and Signal Processing, DOI: 10.1007/978-1-4614-3464-1_1, © Springer Science+Business Media New York 2013

1915, Einstein published an extension of relativity by including gravitation denominated as the general theory of relativity. Some predictions of these path-breaking theories have been confirmed. Nevertheless, he received the Nobel Prize in 1921 for the explanation of the photoelectric effect, a confirmation of the particle nature of light leading to particle-wave duality, and eventually to quantum physics, another research field that saw great development during the twentieth century.

There has been another leap from Einstein's time to the present. Even though research in physics is multidimensional and growing all over the world rapidly, making identification of major changes difficult, microelectronics might be considered as the new leap. The world has changed enormously with the advent of microelectronic technology. Computers and mobiles go with us wherever we go, and a continuous connection to the world is available for everyone.

1.1 Optical Communications

Within the revolution of microelectronics, optical transmission is the key to create broadband communication networks. The increase in users and the capacity per user are only fulfilled by this kind of transmission. In this section, a historical introduction, a brief discussion about emerging applications, a comparison between different communication links, and an overview of an optical link are included.

1.1.1 A Look at History

Rudimentary communications systems based on light were used in antiquity, such as fire beacons and smoke signals. Despite the very limited information capacity of this technique when compared to the information carried by messengers, two advantages must be highlighted: the message could reach several receivers simultaneously and the speed of the transmission is unbeatable. Therefore, this method was preferred for alarm signals, especially in periods of war. For example, the Greek tragedian Aeschylus describes how the message of the fall of Troy (1184 BC) was sent by fire signals via an unbroken line of beacon-fires from Asia Minor to Mycenae (Greece) covering a distance of 600 km in the Oresteia trilogy (458 BC).

By these systems, only a limited number of predetermined messages could be transmitted because of the lack of a transmission code. In "The Histories", the Greek historian Polybius describes the first known telegraph developed by Aeneas. Based on a water level, the entire Greek alphabet could be transmitted by fire signals using a two-digit, five-level code. By means of torches, a protocol was established to start and stop the emptying of identical receptacles simultaneously. Thus, the level of water corresponded to a predetermined message.

There were no significant advances for transmission based on light until the French revolution in the eighteenth century. The civilian Claude Chappe, a former priest, invented a mechanical–optical telegraph. It consisted of a column with a movable crosswise beam and two arms. Each arm had seven positions, and the crosswise beam had four more, permitting a 196-combination code. The equipment acting as repeater stations stood on rooftops of towers located approximately 10–25 km apart. The first telegraph line of this sort was put into operation in 1794 between Lille and Paris. It consisted of 22 stations over 240 km, completing a message transfer in only 2–6 min, whereas riding couriers would require 30 h to perform the same task. Other lines were built, including a line from Paris to Toulon, 765 km via 120 stations, connecting the capital of France with the Mediterranean Sea, and another extending the line via Lille to Brussels in 1803 and Amsterdam in 1810.

The era of electrical communication began in 1838 through the invention of the electrical telegraph by Samuel Morse. In contrast to the optical telegraph that was not suitable for transmission at night even with the lamps attached to the arms, the new transmission method could be used at night and at an increased data rate of about 10 bit per second. The first successful transoceanic telegraph cable between the United States and Europe was put into service in 1865. Later, the telephone was invented by Alexander Graham Bell in 1876 and radio communication was made possible by the contributions of Maxwell (1873), Hertz (1887), and Marconi (1895). Electrical communication systems have evolved considerably since then and bit rates of hundreds of Mb/s have been achieved. However, the distance between the repeaters for a high-speed system is rather small (~1 km), which makes it relatively expensive to operate.

The interest in using light as carrier of communication is reflected in one anecdote. Alexander Graham Bell also invented the photophone in 1880. He considered this discovery greater than his previous one, the telephone. In the photophone, vibrations in the voice from the emitter caused movements in a mirror. Then, light reflected on this mirror excited a selenium crystal to vibrate in the receiver. Although the invention was tested successfully, it failed at night, in rain, or while passing through an obstacle.

All the problems in optical transmission are related to the lack of a transmission channel. However, scientists have realized that the bandwidth–length product of the transmission system can be enhanced if optical waves were used as the carriers. Unfortunately, neither a coherent optical source nor a suitable transmission medium was available until 1970. Optical fibers were suggested as the transmission medium and the advent of the laser in the early 1960s solved the first problem. However, despite the introduction of cladding by Abraham van Heel, the attenuation of the purest glass produced was too high (1000 dB/km) for long-distance transmission.

In 1966, Charles Kao and George Hockam published an article Kao and Hockham (1966) that predicts the feasibility of this kind of communication if the transmission loss was reduced to < 20 dB/km. Moreover, they proved that there were no fundamental technicalities that would prevent this loss from being

achieved. It represents such a milestone that Charles Kuen Kao was awarded the
Nobel Prize in Physics in the year 2009 "for groundbreaking achievements con-
cerning the transmission of light in fibers for optical communication." He shared
this honor jointly with Willard S. Boyle and George E. Smith "for the invention of
an imaging semiconductor circuit—the CCD sensor".[1] Four years later, in 1970,
the goal of reducing transmission loss to < 20 dB/km was achieved by Robert
Maurer, Donald Keck, and Peter Schultz of the Corning Glass Corporation. The
attenuation was reduced by doping the fiber with titanium. In 1980s, the attenuation
loss was reduced further thereby defining the first generation of long-haul optical
links that target a superior distance between the repeaters than its electrical counterpart. The first international undersea link, from England to Belgium, was
installed in 1986. Two years later, the first intercontinental optical fiber system,
TAT-8 between the United States and Europe, was put into service with a total
spanned distance of 6,100 km and a distance between the repeaters of about 40 km.

Table 1.1 summarizes the evolution of long-haul optical communications.
Every generation is characterized by a considerable increase in the bit rate–
distance product. Second and third generations were based on the development of
lasers and detectors operating at longer wavelengths, minimizing the attenuation
and the dispersion of the fiber. Once minimal losses were achieved, optical
amplifiers were introduced to regenerate the signal, extending the electrical
repeater space hugely. At the same time, bit rate was increased by wavelength-
division multiplexing (WDM). Both improvements represent the fourth generation. The era of terabit communications systems has truly arrived with the fifth
generation (Graydon 2002), which is based on dense wavelength-division multiplexing (DWDM), forward error correction, distributed Raman amplification and
solitons (transmitted pulses that are not degraded due to the compensation between
dispersion and nonlinearity). In 2009, an article on transmission was published
Salsi et al. (2009), which targeted a bit-rate–distance product exceeding 100 Pb/s
over 155 channels by digital coherent detection and phase-shift keying (Gnauck
and Winzer 2005). The revolution in communications showing the crossover point
to optical technology is illustrated in Fig. 1.1.

1.1.2 Emerging Applications

Historically, as only optical links satisfy the demands for bit rate–distance product
they have been traditionally exploited for long-haul communications. The considerable cost of the optical system is shared among a large number of potential
users. If a cost-effective solution is achieved, a wide range of applications could be
covered. In the following sections, more possibilities that are promising are
discussed.

---

[1] http://nobelprize.org/nobel_prizes/physics/laureates/2009/

Table 1.1 Generations of optical fiber for long-haul applications (Agrawal 1997; Graydon 2004; Salsi et al. 2009)

Gen.	Year	Bit rate R_b	Distance D (km)	$R_b \cdot$ D product	λ (μm)	Achievements/properties
1st	1980	45 Mb/s	10	150 Mb/s · km	0.8	Superior repeater space than electrical communication Graded-index fiber
2nd	1987	1.7 Gb/s	50	85 Gb/s · km	1.3	Single-mode fiber Minimum dispersion Fiber loss < 1 dB/km
3rd	1990	2.5 Gb/s	70	175 Gb/s · km	1.55	Minimal loss (< 0.2 dB/km) Dispersion-shifted fiber
4th	1996	5 Gb/s	11,300	56.5 Tb/s · km	1.55	Erbium-doped optical amplifiers
	2000	100 Gb/s	9,000	900 Tb/s · km	1.55	Wavelength-division multiplexing (WDM)
5th	2002	1.28 Tb/s	4,000	5120 Tb/s· km	1.55	Solitons Distributed Raman amplification Forward error correction Dense wavelength-division multiplexing (DWDM)
	2009	15.5 Tb/s	7,200	112 Pb/s · km	1.55	Phase-shift keying (PSK) Digital coherent detection

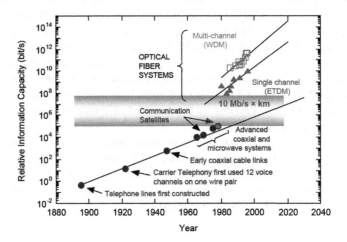

Fig. 1.1 Evolution of communication technologies (MIT Microphotonics Center Industry Consortium 2005)

1.1.2.1 Short-Haul Networks

Currently, bandwidth demands for short-distance communications are increasing exponentially (Ziemann et al. 2008). Thus, the conversion from electrical to optical transmission, as for long-haul communications, could be more beneficial than the improvements in transmission through cooper wires. Despite the advantages, the definition of beneficial is purely based on cost.

To make short-distance fiber communication affordable, the industry has developed low-cost solutions such as high-bandwidth multimode fibers and 850 nm transceivers. Eventually, fiber optics will spread in local area networks (LANs) and new systems like fiber-to-the-home (FTTH) will be commercially available (Brillant 2008). Home networking is commencing; even as the required bit rate in FTTH applications is 100 Mb/s over 50 m, the goal set by some telecom operators for home networking is to achieve speeds of several Gb/s.

1.1.2.2 In-Car Fiber-Optic Networks

Optoelectronic systems have also become increasingly attractive for communication inside cars (Freeman et al. 2004). To connect the ever-increasing number of in-car electrical devices, plastic optical fiber (POF) is used. In addition to the already mentioned benefits, POF networks provide a low-cost solution with an ease of handling and installation, compared to glass optical fiber.

Several protocols are employed, which can be differentiated in two main types: multimedia and security, where the priority is speed and total reliability, respectively. In 1998, an international consortium of car manufacturers and suppliers set up an open standard, the Media Oriented System Transport (MOST), which can connect the radio, the CD/DVD player, the navigation system, a Bluetooth interface, telephones, games consoles, and a voice-recognition system inside a car. Another multimedia protocol is the IDB-1394, the automotive version of IEEE-1394. Security protocols have been developed to communicate with the rapidly growing number of sensors, actuators, and electronic control units within cars. For instance, BMW's Series 7 models implement byteflight for control of the car's air-bar systems, whereas FlexRay is the standard that will be used in the next-generation drive-by-wire systems.

All currently in-car optical data bus systems are basically the same components: standard poly-methyl methacrylate (PMMA) POF, red (650 nm) emitting LEDs, and large silicon photoreceivers. Recently, 1 mm step index-POF (SI-POF) has been standardized as A4a.2 and is being used massively in the automotive sector (25 Mb/s) and industrial automation (100 Mb/s).

1.1.2.3 High-Speed Optical Interconnects

The highest number of transistors per unit of area within a single chip has been doubled every 18–24 months according to Moore's prediction (Moore et al. 1965). Therefore, the processing speed has also been considerably enhanced and interconnects are becoming a bottleneck. Moreover, over the next decade, the bandwidth of interconnects inside a computer is expected to increase by an order of magnitude, from 1 to 10 GHz, requiring some kind of internal optical data-bus to overcome this problem.

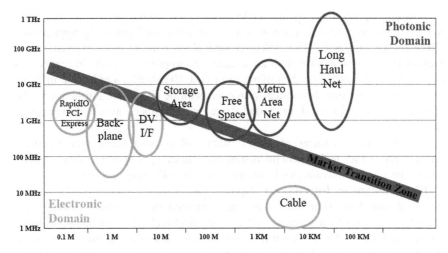

Fig. 1.2 Bandwidth—distance market map (MIT Microphotonics Center Industry Consortium 2005)

According to experts in the field, optics could be playing a role in board-to-board links soon although it will take some years before optical interconnects will be employed for chip-to-chip communication (Savage 2002; Graydon et al. 2004). Optical interconnects has been suggested even to connect the subsystems within a single chip, but this approach is under intense discussion.

1.1.3 Comparison Between Communication Links

Optical communications systems compete with electrical wire data transmission and wireless communications. The superior performance of optical communications, defined by bandwidth–distance product, has motivated a gradual conversion into photonic domain crossing the market transition zone to attain higher transmission speed, as illustrated in Fig. 1.2.

Very high performance fiber optic systems are relatively expensive; therefore, they are limited for applications where the cost is shared among a certain amount of users. For short-reach applications, where several possibilities (electrical, optical, and wireless) of solving the same demand are offered in the market, two variables represent the breakpoint for the choice: cost and mobility.

The obvious advantage of wireless communications is the mobility. Particular users mandate devices permanently connected. Mobile and WiFi communications offers this possibility, although they cannot be considered a counterpart of optical communications. Neither mobile nor WiFi targets gigabit communications and their emitters and receivers must be connected by some communication system. In particular, the optical and WiFi combination is a very attractive idea.

Low-cost optical links are available and are very competitive with wire technology, spreading optical transmission to the mentioned emerging applications (Hermans and Steyaert 2007). Although optical systems are more expensive than their electrical counterparts, their advantages and improvements could represent a more important factor than the difference of cost. Eventually, a reduction in terms of cost and design time for optical systems cannot be dismissed. Furthermore, the unlimited supply of sand to manufacture optical fibers avoids future price increases.

The dielectric character of optical fibers offers advantages over electrical counterparts, which make them interesting for certain types of applications. A brief discussion of these advantages is presented in the following sections.

Electromagnetic Interference Optical carrier has no charge whereas an electrical one has. That little difference between optical and electrical data transmission leads to a great advantage. An electrical wire conducting a high-speed signal may act as a transmitting antenna and radiate noise, possibly causing interference-related problems in neighboring circuits. On the other hand, electromagnetic noise from the outside world may disturb data transmission. This issue has been traditionally solved by using heavy shielded cables or balanced lines with differential drivers and receivers. Optical fiber systems directly avoid electromagnetic interference (EMI), being an alternative for transmission through areas with an important electromagnetic pollution. For instance, they are applicable in industrial environments, where a considerable electromagnetic noise is caused by heavy industrial machinery. In the other direction, optical fibers are very attractive for applications with restrictions on the tolerable electromagnetic radiation, easing the noise immunity requirements for the subcircuits in large systems. The lack of emission also makes the optical fiber unbeatable for secure data or voice transmission, ensuring confidentiality.

Galvanic Separation Ground potential may vary depending on the location. Thus, when an electrical signal is transmitted over a certain distance, ground loops need being solved. They manifest as currents flowing through the shield or ground wire of the interconnecting cable because of differences in the local reference potentials. Balanced lines, differential circuits, and optocouplers are commonly used to solve these issues. Optical transmission offers an ideal solution as it provides an inherent isolated data path.

Security In some areas, safety must be maximized. Optical transmission offers an improvement, as no electrical current is conveyed. As a clear example, in an area with volatile chemicals any spark generated may lead to a catastrophe. In addition, fiber cables are not affected by corrosion, being well suited for corrosive environments. Furthermore, if the communication system must be manipulated, there is no possibility of electrocution or short circuit hazard.

Weight An optical fiber is much lighter than an electrical wire for a given length. The light weight and small size make the fiber interesting for specific applications such as "fly by wire" technique in airplanes. The lightness of the fiber also enables a weightless to carry communication system, which is, for example, of interest in military tactical operations.

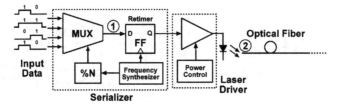

Fig. 1.3 Complete diagram of transmitter system

Environment Electrical wires and optical fibers are manufactured from different raw materials. Glass for optical fibers is obtained from sand without affecting the environment in contrast with the extraction of copper. Therefore, based on ecological considerations, optical links must be chosen as it helps conservation of earth's resources.

1.1.4 Optical Link Architecture

This section presents an overview of the whole transmission system (Razavi 2003), from electrical to optical conversion and vice versa. The system consists of the transmitter, the transmission channel, and the receiver (Fig. 1.3). A similar subdivision can be easily observed in the transmitter and receiver, taking into account the digital and analog signal processing and the conversion between the electrical and optical signals.

The transmitter is subdivided as a serializer, a driver, and a light emitter. Several synchronous digital signals are multiplexed into a transmitted digital signal by the serializer. It consists of a multiplexer (MUX) and a frequency synthesizer, based on a phased-locked loop (PLL). Non-idealities caused by the MUX are absorbed by the retimer. The laser driver must provide the proper modulated current to the light emitter and usually incorporates power control. A laser is commonly used as the light emitter because of the higher output power and more spectral purity. For low-cost solutions, a light-emitting diode (LED) can be adopted as the light emitter.

The receiver consists of a photodiode, a front-end, and a deserializer, as illustrated in Fig. 1.4. The photodiode converts the transmitted optical power into a current, which can be electronically processed by the front-end to provide a signal with sufficient quality. The front-end is basically formed by a transimpedance amplifier and a post-amplifier, which converts the photocurrent into a voltage and boosts such a voltage swing to logical levels, adequate for subsequent digital circuitry, respectively. A post-amplifier may consist of a simpler cascade amplifier chain, denominated as the limiting amplifier, or present additional circuitry to control the gain, denominated as automatic gain control (AGC) amplifier. This latter implementation offers linear operation for a wider range of input signal amplitude, permitting analog signal

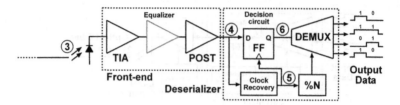

Fig. 1.4 Complete diagram of receiver system

processing to be performed on the output signal. The equalizer is widely employed when the channel or photodiode causes band-limited effect, for instance, in POF channels or with complementary metal-oxide semiconductor (CMOS)–integrated photodiode. The deserializer must target two main functions: clock and data recovery and demultiplexing. First, from the received signal, it must recover the associated clock signal. Then, the received signal is converted to a digital signal by deciding between the two possible states indicated by the recovered clock. Finally, the digital signal is demultiplexed.

This diagram can be generalized with optical wavelength multiplexing, including an array of transmitters and receivers (Muller and Leblebici 2007), one pair for each wavelength. Therefore, the required data rate for each transmitter/ receiver is relaxed but the power consumption of the system is increased.

To summarize, Fig. 1.5 shows the typical signal along the transmission path. Corresponding places between these signals and Figs. 1.3 and 1.4 are indicated by numbers within circles. First, the serialized digital data is converted to a modulated optical signal by the driver. This signal is attenuated and dispersed along the optical fiber, and so, it must be processed by the front-end to recover the original data. The electronic noise from the front-end represents the main contribution to the output noise. Clock and data recovery circuit regenerates the clock and the transmitted data.

1.2 CMOS Technology

The bipolar transistor was developed in 1947 by John Bardeen, Walter Brattain, and William Shockley, who were awarded the Nobel Prize in 1956 "for their researches on semiconductors and their discovery of the transistor effect".[2] John Bardeen is the only laureate to win the prize twice. He was also awarded in 1972, shared with Leon Cooper and John Schrieffer, "for their jointly developed theory of superconductivity, usually called the BCS-theory".[3] CMOS technology was patented by Frank Wanlass in 1967, although the basic principle of a kind of

---

[2] http://nobelprize.org/nobel_prizes/physics/laureates/1956/

[3] http://nobelprize.org/nobel_prizes/physics/laureates/1972/

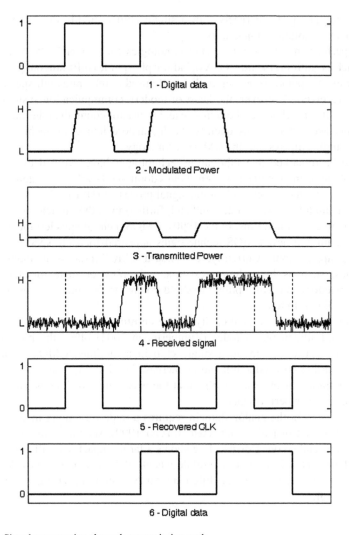

Fig. 1.5 Signal propagation through transmission path

field-effect transistor (FET) was described on a patent filed by physicist Julius Edgar Lilienfeld in Canada in 1925. This is the starting point of a technology that changes everything.

Although the system topologies of Figs. 1.3 and 1.4 have not changed much over the past several decades, the design of its building blocks and the levels of integration have (Razavi 2003). Motivated by the evolution and affordability of integrated circuit (IC) technologies as well as the demand for higher performance, optical links traditionally consist of several single-chip blocks integrated in different technologies, exploiting the main advantages of each one. A typical partitioning is indicated both in Figs. 1.3 and 1.4 by dashed boxes. Nevertheless, the full integration of the receiver in a single chip is highly desirable, and for a cost-effective solution, mandatory.

High performance (and high cost) technologies (Säckinger 2005; Singh et al. 2004), such as gallium arsenide (GaAs), indium phosphide (InP), silicon–germanium (SiGe) or even bipolar, have been traditionally used to integrate high-speed optical front-end receivers. However, the heart of modern communication systems is the digital core. It is much larger, both in size and transistor count, than the surrounding analog interface circuits. Consequently, the digital core dictates the technology to be used in a single chip system: the CMOS technology.

It must be remarked that fabrication technologies have evolved to satisfy the demand of the IC market that has been always dominated by digital systems, such as microprocessors, memories, and digital signal processor (DSP). CMOS technology is the most suitable for such a demand and, furthermore, the fabrication process is unbeatable in economic terms. In the 1990s, CMOS technologies left the exclusive digital domain and new CMOS technologies appeared focusing special attention on analog applications. Currently, a standard CMOS process includes analog capabilities, such as high-ohmic resistors, poly–poly or metal–insulator-metal (MIM) capacitors and multioption threshold voltage to facilitate the integration of analog and digital functions.

The development of vertical cavity source emitting lasers (VCSEL) (Gulden et al. 2001) and silicon photodiodes (Zimmermann 2004) offers the possibility to integrate the entire system on a single chip in a modified CMOS technology, attaining a further cost reduction and an increase in reliability. However, the large length penetration of photons in silicon at the wavelength of interest shows a big impact on receiver performances.

CMOS design in the analog domain requires a trade-off among many different factors that are intrinsically connected (Razavi 1999), such as gain, noise, bandwidth, power, and so on. Downscaling shows a large impact over all these factors (Abou-Allam 2000), leading to several drawbacks that must be faced by designers. Above all, the choice between the different CMOS technologies is based on a cost–speed trade-off.

1.3 State of the Art

The state of the art can be clearly split into three subtopics: very high-speed optical transmission per channel, full integration of the optical receiver in the same technology, and cost optimization for single user applications.

The growth in Internet data traffic and computation power in recent years has increased the demand for more speed in almost all communication systems. Expensive technologies, such as InP (Shigematsu et al. 2001) or SiGe (Reinhold et al. 2001), are employed to target 40 and 100 Gb/s standards. Meanwhile, CMOS technology in nanometer scale has demonstrated promising results (Kromer et al. 2004; Liao and Liu 2007).

The cost and integration advantages of CMOS technology have motivated extensive work on submicron CMOS design for optical applications. In the last few years, optical receivers at 2.5 Gb/s (Mitran et al. 2002), transimpedance amplifiers with inductorless broadband techniques (Tsai and Chen 2007) and automatic gain control amplifiers with linear in dB gain control (Wu et al. 2004) have been published. In 2002, 0.18 μm CMOS serializers (Green et al. 2002) and deserializers (Cao et al. 2002) operating at 10 Gb/s were reported. To target such a bit rate for analog front-end in submicron CMOS technology, inductors can be implemented. A transimpedance amplifier (Tao and Berroth 2003), a limiting amplifier (Galal and Razavi 2003), and an optical receiver (Chen et al. 2005) at 10 Gb/s and even a limiting amplifier (Galal and Razavi 2004) at 40 Gb/s have been published in the previous years. However, integrated inductors substantially increase the chip area.

A further step in the integration of the full transmission system in CMOS technology can be obtained by implementing the photodiode on a silicon substrate. Silicon photodiodes are not suitable for wavelengths >1 μm and present a frequency response mandating of an equalizer to achieve multi-gigabit transmission. In the literature, there are some examples targeting 1.2 Gb/s (Hermans et al. 2006), 3 Gb/s (Radovanovic et al. 2005), 3.125 Gb/s (Chen et al. 2007), 4.5 Gb/s (Tavernier and Steyaert 2008), and 5.5 Gb/s (Tavernier and Steyaert 2010a).

To reduce the cost of the optical link, POF replaces glass optical fiber. Because of the large core of this kind of fiber, the optical receiver must deal with a large photodiode in addition to the more restrictive bandwidth–length product of the fiber itself. High speed transmission over POF has been reported in the literature, targeting 1 Gb/s (Zerna et al. 2009), 3 Gb/s (Dong and Martin 2010), and 800 Mb/s with an integrated photodiode (Tavernier and Steyaert 2010b).

This book focuses on the CMOS design of the front-end receiver, motivated by the integration of the whole communication system on the same technology (Ingels and Steyaert 2004). The front-end receiver consists of a transimpedance amplifier that converts the input photocurrent into a voltage, a post-amplifier to boost the signal up to a proper level and an equalizer depending on the application.

1.4 Outline of the Work

The overall aim of this work is to develop architectures of building blocks for optical receivers suitable for gigabit standards. Therefore, the analog front-end must offer high-speed operation and a sufficient output voltage signal level to be able to recover the data signal. In addition, noise performance must be optimized to improve the sensitivity of the receiver. Finally, some techniques to improve the dynamic range will be explored.

The optical front-end receiver will be implemented in advanced submicron CMOS technologies to facilitate the integration on a single chip joined to the subsequent digital circuitry, leading to a cost-effective solution. This raises several critical issues in the design of all building blocks of the analog front-end.

Although it is not the predominant methodology for microelectronic design, application-specific integrated circuits (ASICs) provide unquestionable advantages owing to the integration of analog and digital functions over the same substrate: smaller size, greater packing densities, high operation speed, fewer connection failures, lower parasitic effects, reduced cost, lower power consumption, and more versatility in the design. Because of the presented advantages, ASICs are the preferable choice for a wide range of applications, including telecommunications.

Overall, the main aspects developed throughout this book are the following:

- First, theoretical fundamentals of the digital signal will be studied, defining the main parameters, such as bit rate and disparity, analyzing data in terms of frequency domain and introducing the eye diagram, the pseudorandom binary signal (PRBS) signal and typical code formats.

- Next, an overview of a complete electro-optical system is briefly introduced, leading to the definition of three key parameters: bit error rate of transmission, and sensitivity and dynamic range of the optical receiver. In addition, bit error rate is related to noise performance by a Gaussian model, and sensitivity penalties because of some undesirable effects, such as extinction ratio and decision offset, are analyzed.

- As revealed from a review of the literature, the analog front-end is usually divided into two main building blocks: a transimpedance amplifier and a post-amplifier. The state of the art of both blocks was analyzed, exhibiting the advantages and drawbacks of typical architectures by comparing the previously published works, and revealing the most suitable structure and the main performances to achieve.

- The most important block of the optical receiver, the transimpedance amplifier, will be carefully designed. The optimization of two key parameters—sensitivity and dynamic range—of the optical receiver, which are basically restricted by this block, will be studied. A highly sensitive receiver targets sensitivity as low as -30 dBm, although the requirement depends on the application. The dynamic range of the TIA should cover an input signal level of 1 mA peak to peak, which is a typical order of magnitude for the highest current provided by a photodiode.

- Requirements for noise performance are relaxed for the post-amplifier. In contrast, broadband techniques were fundamental to achieve the more demanding speed requirements. The bandwidth of the post-amplifier should be around or superior to the bit rate, while a high gain (30 dB) is required, leading to a very high gain–bandwidth product.

- Instead of the conventional limiting amplifier architecture, an automatic gain control amplifier is proposed to implement the post-amplifier, which leads to some advantages but more requirements. In addition, there will be offset compensation. The time constant of both control loops must be around 1 μs to target the specs of communication standards.

- A low-cost CMOS receiver, based on the proposed building blocks, for short-reach optical communications is presented. The goal is a gigabit data rate transmission over plastic optical fiber for a few meters reach, which mandates to include an equalizer in the receiver chain to compensate the band-limited effect of such a fiber.

- Because of the broad frequency range, test setups, so as not to degrade the real performance of the device, are a critical issue in this work. The considered approaches for measurements are presented, such as on-board, on-wafer, and electrical characterization for transimpedance amplifiers.
- To conclude, the main fulfilled objectives of the book will be summarized and a proposal for future research directions will be drawn.

To cover all this aspects, this book is set out in six chapters; the first one includes this introduction and the last chapter presents conclusions of the whole work. In all other chapters, a section is reserved at the end for conclusions drawn for that chapter and the bibliography employed. To facilitate the reading of the book, contents, figure and table index, and the list of symbols and abbreviations employed are offered at the beginning. Finally, some appendixes are added.

This introductory chapter begins by exploring the history of physics, and in particular, of optical transmission. Then, the characteristics of such a transmission are analyzed, showing improvements and advantages over other possibilities. Thanks to these advantages, the range of applications covered by optical communications is increasing according to the requirements of transmission. In addition, a brief discussion of the chosen microelectronic technology is made. In the end, the outline of the work is offered showing the aims to be achieved.

The Chap. 2 covers the theoretical fundamentals of optical transmission. It is divided into three sections. First, the data signal is analyzed, explaining different binary formats, codes, its properties, and the typical eye diagram representation. Next, the fundamentals of optical fibers and an overview of a complete optical system are presented. Finally, a detailed definition of key parameters (bit error rate, sensitivity, and dynamic range) is provided.

The core of this work is offered in Chaps. 3–5. The design and verification of the two building blocks of a front-end receiver are presented in the signal path order, that is, first transimpedance amplifier (TIA), and then, post-amplifier. Simulation results are included to illustrate the design stage. A new compression technique for the TIA and an automatic gain control (AGC) loop for the post-amplifier (PA) are implemented to extend the input dynamic range of the whole receiver. Multistage structure and broadband techniques must be implemented in a PA to fulfill the gain-bandwidth requirements. Both blocks are verified with frequency and time-domain measurements.

The fifth chapter presents a complete prototype of an optical receiver for short-reach applications, including building blocks designed in previous chapters. A cost-effective solution for these applications is the POF, whose limitations are introduced in this chapter. To achieve gigabit data rates for such a fiber, equalization is mandatory. Therefore, the proposed architecture includes an equalizer in the receiver chain to fulfill gigabit transmission.

To conclude, the final chapter summarizes the main conclusions drawn from the experimental results obtained in the previous chapters. In addition, the research directions that could be further studied in the future are analyzed.

At the end of this book, several appendixes are included. A brief explanation of the measurement considerations including S parameters, calibration, and de-embedding is offered in the first appendix. The PRBS generator and characteristics of this kind of signals are introduced in the second appendix. The last one summarizes the process features and parameters of the considered 180 and 90 nm CMOS integrating technologies. Finally, the index of this book is added.

References

Abou-Allam E, Manku T, Ting M, Obrecht M (2000) Impact of technology scaling on CMOS RF devices and circuits. IEEE custom integrated circuits conference, pp 361–364

Agrawal GP (1997) Fiber-optic communications systems, 2nd edn. Wiley, New York

Brillant A. (2008) Ditigal and analog fiber optic communications for CATV and FTTx applications. SPIE and Wiley

Cao J et al (2002) OC-192 Receiver in standard 0.18 μm CMOS. ISSCC digest of technical papers, pp 187–188

Chen W-Z, Cheng Y-L, Lin D-S (2005) A 1.8 V 10 Gb/s fully integrated CMOS optical receiver analog front-end. IEEE J Solid-State Circ 40(6):1388–1396

Chen WZ, Huang SH, Wu GW, Liu CC, Huang YT, Chin CF, Chang WH and Juang YZ (2007) A 3.125 Gbps CMOS fully integrated optical receiver with adaptative analog equalizer. In: Proceedings of the 2007 IEEE Asian solid-state circuits conference, pp 396–399

Dong Y and Martin K (2010) Analog front-end for a 3 Gb/s POF receiver. In: Proceedings of the 2010 IEEE international symposium on circuits and systems, pp 197–200

Freeman T (2004) Plastic optical fibre tackles automotive requirements. Fibresystems Europe/ LIGHTWAVE Europe, pp 14–16

Galal S, Razavi B (2003) 10 Gb/s limiting amplifier and laser/modulator driver in 0.18 μm CMOS technology. IEEE J Solid-State Circ 38(12):2138–2146

Galal S, Razavi B (2004) 40 Gb/s amplifier and ESD protection circuit in 0.18 um CMOS technology. IEEE J Solid-State Circ 39(12):2389–2396

Gnauck AH, Winzer PJ (2005) Optical phase-shift-keyed transmission. J Lightwave Technol 23(1):115–130

Graydon O (2002) "The Terabit Challenge", Optics & Laser Europe, pp 31–32

Graydon O (2004) Photonics unlocks chip bandwidth bottleneck. Opt Laser Europe pp 25–27

Green MM et al (2002) OC-192 transmitter in standard 0.18 μm CMOS. ISSCC digest of technical papers, pp 186–187

Gulden KH et al (2001) VCSEL arrays for high speed optical links. Annual techl digest gallium arsenide integrated circuit symposium, pp 53–56

Hermans C and Steyaert M (2007) Broadband opto-electrical receivers in standard CMOS. Analog circuits signal process. Springer, Berlin

Hermans C, Tavernier F and Steyaert M (2006) A gigabit optical receiver with monolithically integrated photodiode in 0.18 μm CMOS. IEEE European solid-state circuits conference, pp 476–479

Ingels M, Steyaert M (2004) Integrated CMOS circuits for optical communications. Advanced microelectronics. Springer, Berlin

Kao KC, Hockham GA (1966) Dielectric fibre surface waveguides for optical frequencies. Proc IEE 113(7):1151–1158

Kromer C et al (2004) A low-power 20 GHz 52 dBΩ transimpedance amplifier in 80 nm CMOS. IEEE J Solid-State Circ 39(6):885–894

Liao C, Liu S (2007) A 40 Gb/s transimpedance-AGC amplifier with 19 dB DR in 90 nm CMOS. IEEE international solid-state circuits conference pp 54–55, 586

MIT Microphotonics Center Industry Consortium (2005) Microphotonics: hardware for the information age. communications technology roadmap

Mitran P, Beaudoin v and El-Gamal MN (2002) A 2.5 Gbit/s CMOS optical receiver frontend. In: Proceedings of the 2002 IEEE international symposium on circuits and systems, vol 5, pp 441–444

Moore GE (1965) Cramming more components onto integrated circuits. Electronics 38(8):114–117

Muller P, Leblebici Y (2007) CMOS multichannel single-chip receivers for multi-gigabit optical data communications. Analog circuits and signal processing. Springer, Berlin

Radovanovic S, Annema AJ, Nauta B (2005) A 3 Gb/s optical detector in standard CMOS for 850 nm optical communication. IEEE J Solid-State Circ 40(8):1706–1717

Razavi B (1999) CMOS technology characterization for analog and RF design. IEEE J Solid-State Circ 34(3):268–276

Razavi B (2003) Design of integrated circuits for optical communications. McGraw-Hill, New York

Reinhold M et al (2001) A fully integrated 40-Gb/s clock and data recovery IC with 1:4 DMUX in SiGe technology. IEEE J Solid-State Circ 36:1937–1945

Säckinger E (2005) Broadband circuits for optical fiber communication. Wiley, Hoboken

Salsi M, Mardoyan H, Tran P, Koebele C, Dutisseuil E, Charlet G, Bigo S (2009) 155x100Gbit/s coherent PDM-QPSK transmission over 7,200 km. In: Proceedings of the 35th European conference on optical communication, pp 1–2

Savage N (2002) "Linking with Light", IEEE Spectrum 32–36

Shigematsu H, Sato M, Suzuki T, Takahashi T, Imanishi K, Hara N, Ohnishi H, Watanabe Y (2001) A 49 GHz preamplifier with a transimpedance gain of 52 dB using InP HEMTs. IEEE J Solid-State Circ 36(9):1309–1313

Singh R, Harame DL and Oprysko MM (2004) Silicon germanium: technology, modeling and design. IEEE Press, Piscataway

Tao R and Berroth M (2003) A 10 Gb/s fully integrated CMOS transimpedance preamplifier. IEEE European solid-state circuits conference, pp 549–552

Tavernier F, Steyaert M (2008) Power efficient 4.5 Gbit/s optical receiver in 130 nm CMOS with integrated photodiode. IEEE European solid-state circuits conference, pp 162–165

Tavernier F, Steyaert M (2010a) A 5.5 Gbit/s optical receiver in 130 nm CMOS with speed-enhanced integrated photodiode. IEEE European solid-state circuits conference, pp 542–545

Tavernier F, Steyaert M (2010b) A high-speed POF receiver with 1 mm integrated photodiode in 180 nm CMOS. 36th European conference and exhibition on optical communication, pp 1–3

Tsai C-M and Chen W-T (2007) A 40 mW 3.5 kΩ 3 Gb/s CMOS differential transimpedance amplifier using negative-impedance compensation. IEEE international solid-state circuits conference

Wu C, Liu C and Liu S (2004) A 2 GHz CMOS variable-gain amplifier with 50 dB linear-in-magnitude controlled gain range for 10 GBase-LX4 ethernet. IEEE international solid-state circuits conference

Zerna C, Sundermeyer J, Tan J, Fiederer A and Verwaal N (2009) Adaptive integrated equalizing techniques for SI-POF home networking links. 18th international conference on plastic optical fibers

Ziemann O, Krauser J, Zamzow P, Daum W (2008) POF handbook: optical short range transmission systems. Springer, Berlin

Zimmermann H (2004) Silicon optoelectronic integrated circuits, advanced microelectronics. Springer, Berlin

Chapter 2
Optical Signal Transmission

In this chapter, the principles of the optical signal transmission will be explored. First, the characteristics of transferred data will be analyzed, focusing on the pseudorandom bit sequence (PRBS), which is the typical signal used to test digital communication prototypes. Second, the fundamentals and the main types of optical fibers will be explained. Next, an overview of the core building blocks of an electro-optical transceiver front-end will be presented, explaining each component and its main requirements. Finally, the main key parameters of optical transmission from receiver's point of view are defined, detailing the Gaussian noise model to determine the sensitivity with respect to the noise performance and the main penalty sources.

2.1 Data Signal

In massive storage systems, such as hard disks, compact discs, and USB sticks, data is digitally stored due to the main advantages of its format: It is perfectly compatible with digital computation and, in particular, its resistance to corruption. To highlight the huge amount of information, according to the International Data Corporation (IDC), the "digital universe"—281 exabytes[1]—exceeded available storage for the first time in 2007 (Gantz et al. 2008).

Focusing on the purpose of this book, digital transmission offers several advantages over analog transmission. The main advantage is the inherent immunity to noise of digital transmission. That is the reason why digital transmission is widely used wherever technology is suitable for such an application. Thus, the intrinsic properties of the storage system—digital format—and advantages of digital transmission are reflected on the characteristics of the data signal.

---

[1] 1 Exabyte (EB) = 10^9 GB = 10^{18} bytes.

F. Aznar et al., *CMOS Receiver Front-ends for Gigabit Short-Range Optical Communications*, Analog Circuits and Signal Processing, DOI: 10.1007/978-1-4614-3464-1_2, © Springer Science+Business Media New York 2013

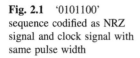

Fig. 2.1 '0101100'
sequence codified as NRZ
signal and clock signal with
same pulse width

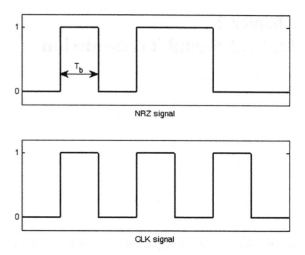

Finally, there is another characteristic of the data signal that must be taken into account from the beginning: It can be treated as a random signal. It is easy to understand because there is no way to predict the information that is being transmitted.

2.1.1 Time Domain

The aforementioned bit sequence can be encoded and modulated in several ways. The simplest one is Non Return to Zero (NRZ) codification and Amplitude-Shift Keying (ASK) modulation, usually denominated as NRZ for simplicity. As illustrated in Fig. 2.1, NRZ signal is formed by a sequence of two possible states— High (H or "1") and Low (L or "0")—with a constant width for each bit (T_b).

The key parameter of the bit sequence is the pulse width, T_b, shown in Fig. 2.1. A digital bit sequence shows three degrees of freedom, namely, the pulse width, the high value, and the low value, but the pulse width is the only one that defines the speed of the transmission. The bit rate R_b—number of pulses transmitted in one second—for an NRZ signal can be expressed as

$$R_b = \frac{1}{T_b} \tag{2.1}$$

It must be noted that for NRZ, each piece of data is transmitted with each pulse, and hence, bit rate and data rate are equivalent. Alternatively, other modulations and codifications for a bit sequence have been proposed (Grigoryan et al. 2003) owing to three main reasons. First, to facilitate data recovery, second, to enhance the amount of data transmitted, and finally, to limit the number of consecutive identical states. Figure 2.2 compares, for the same bit sequence 0011010010 and

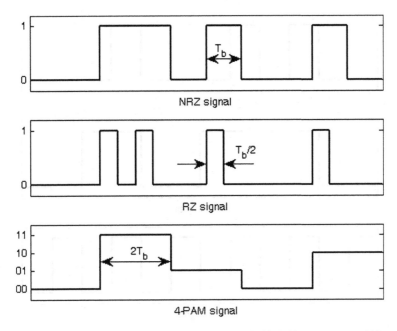

Fig. 2.2 NRZ compared with RZ and 4-PAM for 0011010010 bit sequence and the same data rate

data rate, NRZ signal with a different codification—return to zero (RZ)—and modulation—4-level pulse amplitude modulation (4-PAM)—which provide one of the advantages mentioned earlier.

From Fig. 2.2, it can be seen that RZ signal is formed by an NRZ signal with a redundant "0" between the transmitted bits. Then, the required bit rate to transmit the same data is doubled.

$$R_b|_{RZ} = \frac{2}{T_b} \tag{2.2}$$

Conversely, as shown in Fig. 2.3, RZ signal comes from the composition of an NRZ signal with a clock signal. Thus, it is easy to understand that the data and clock recovery of the transmitted data is facilitated.

To increase the amount of data transmitted within the same width of bit sequence, a multi-level modulation can be used (Pollard 1991). 4-PAM signal transmits twice the information for the same width, when compared with the NRZ signal, or the same information with double width, as shown in Fig. 2.2. For example, to transmit the sequence "0011010010," 4-PAM code uses only five packets "00," "11," "01," "00," and "10," while NRZ requires 10 packets, one for each bit. Subsequently, the relationship between R_b and T_b for 4-PAM can be written as

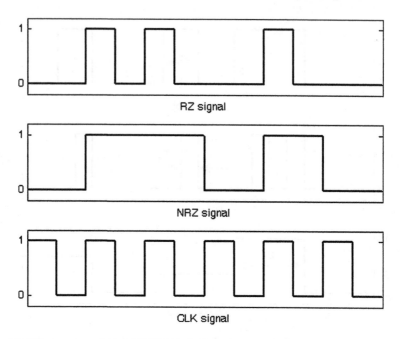

Fig. 2.3 RZ as a composition of NRZ and clock

$$R_b|_{4-PAM} = \frac{1}{2T_b} \qquad (2.3)$$

Thus, this code provides a way to enhance the data transmitted without modifying the pulse width. Ideally, the capacity of the transmission can be further increased by this technique. However, in a real situation, the capacity is limited by the Shannon's theorem (Shannon 1949), which determines the highest data rate that can be transmitted through an analog channel subject to an additive white Gaussian noise.

The definition of the bit rate can be modified by including the following two aspects: Overhead—ratio of bits transmitted which are not data—and multi-level format, resulting in

$$R_b = \frac{1 + \frac{overhead(\%)}{100}}{\log_2 n \cdot T_b} \qquad (2.4)$$

where n is the number of levels, usually a power of 2 to encode a binary word. Thus, the required bit rate for the receiver to transmit the same data is more demanding for a code with overhead and is relaxed for a multi-level format. On comparing (2.4) and (2.2), RZ signal can be seen as a code with 100 % overhead.

Table 2.1 Comparison of several line codes

Code	Overhead (%)	Disparity (bits)	Maximum run length (bits)	References
6b/8b	33	0 @ 8	6	Widmer (2005)
8b/10b	25	±2 @ > 20	5	Widmer and Franaszek (1983)
16b/18b	13	±4 @ > 36	7	Widmer (1999)
64b/66b	3	–	65	Walker and Dugan (2000)

All the previously mentioned formats show the same drawback; the data signal might consist of a constant level—run—during a certain period of time, for example, if several consecutive "0" bits are transmitted. To avoid this possibility, some line codes can be implemented. They consist of increasing the number of pulses transmitted to avoid long runs, minimize the disparity—difference between "1" bits and "0" bits transmitted—and achieve a DC-balance code. They are usually denominated as xb/yb, where y pulses are used to transmit x bits. Table 2.1 shows a comparison among several proposed line codes, indicating the overhead and the maximum disparity offered by the line code.

In the end, it is the communications standard that imposes the format. From here on (unless anything is specified), NRZ signal is used to facilitate the simulations and the test of the prototypes, and compare the results with the state of the art.

2.1.2 Frequency Domain

Another interesting point of view of a random binary sequence is its representation in the frequency domain. This would help us to understand some properties of the transmitted signal.

This brief study starts from the simplest piece of a random binary sequence: A pulse function, which is defined as

$$\text{pulse}(x) = \begin{cases} 1 & \text{if } |x| \leq \dfrac{1}{2} \\ 0 & \text{if } |x| > \dfrac{1}{2} \end{cases} \tag{2.5}$$

Such a pulse(x) function and its Laplace transform sinc(x) are represented in Fig. 2.4, where sinc(x) function is defined as

$$\text{sinc}(x) = \frac{\sin(\pi x)}{\pi x} \tag{2.6}$$

corresponding to a normalized definition.[2] It can be seen that almost the whole spectrum is contained in the pulse function. This is due to the fact that the pulse

---

[2] Its integral over all x is equal to 1.

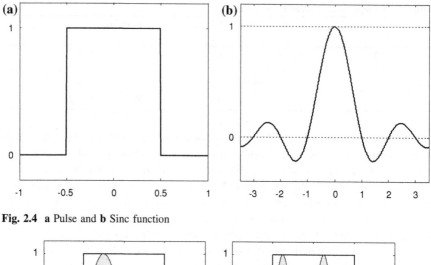

Fig. 2.4 a Pulse and **b** Sinc function

Fig. 2.5 Example of sinusoidal signals (*first* and *second*) which are not included in pulse function spectrum as the positive (*light grey*) and negative (*dark grey*) area cancels

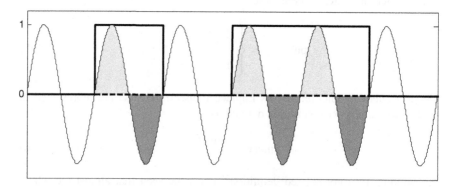

Fig. 2.6 Example of sinusoidal signal which is not included in NRZ spectrum as positive (*light grey*) and negative (*dark grey*) area included in pulses cancels

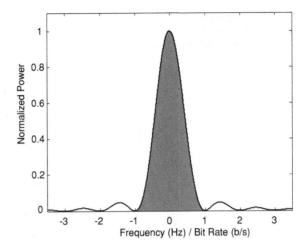

Fig. 2.7 NRZ frequency spectrum. *Grey* area represents the 90 % of the total power transmitted

function is not periodic. Only the sinusoidal signal, whose period is an integer n times a fraction of the width of the pulse signal, is not a component of such a signal. This effect is illustrated in Fig. 2.5 for the two first non-components of the pulse function.

As shown in Fig. 2.6, this cancellation can be also applied to an NRZ signal, because it is a pulse sequence. This means that there is no component of NRZ signal with frequency equal to its bit rate. It could appear strange, but one can notice that the fastest sequence ("...101010...") corresponds to a square signal whose first harmonic frequency is half of the bit rate. According to the definition of pulse width and bit rate of an NRZ signal, the non-frequency components f_n are located in

$$T = \frac{T_b}{n} \Rightarrow f_n = nR_b \qquad (2.7)$$

Accordingly, it can be demonstrated that the frequency spectrum for a random sequence of pulses (NRZ signal) is the same as that for a single pulse (Couch 2007). Therefore, the spectrum in terms of power of the NRZ signal is proportional to $sinc^2$, as illustrated in Fig. 2.7. The gray area, which is the transmitted power for frequencies below the bit rate, represents more than 90 % of the total power transmitted.

This result can be extrapolated to deduce that the 4-PAM spectrum appears like an NRZ spectrum with half bit rate for the same data rate. Similarly, as the RZ signal can be seen as a composition of an NRZ signal and a clock (square signal), the RZ frequency spectrum can be obtained by adding the NRZ spectrum for each harmonic of the square signal, depending on its location and its weight. Figure 2.8 compares the NRZ and 4-PAM spectra with the obtained spectrum for RZ up to 11th harmonic of the square signal. Thus, RZ spectrum is similar to a NRZ spectrum with double bit rate.

Fig. 2.8 Comparison
between NRZ, RZ and 4-
PAM frequency spectrum

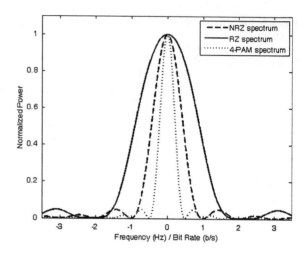

2.1.3 Pseudorandom Bit Sequence

The characterization of the receiver in time domain requires the generation of a
random digital signal. Software and hardware sequence generators are based on
recurrence rules, so the generated signal is not strictly random. A signal as random
as possible must be generated to test the receiver properly.

A bit sequence a_j can be expressed as

$$a_j = \begin{cases} 1 \\ 0 \end{cases} \text{ for } j = 1, 2, \ldots, N \tag{2.8}$$

where N is the length of the sequence. If there is m ones (and $N - m$ zeros), two
parameters can be defined, the duty cycle c and the disparity D, respectively

$$c = \frac{m - 1}{N - 1} \tag{2.9}$$

$$D = \frac{2m - N}{N} \tag{2.10}$$

Duty cycle reflects a normalization of the number of ones, and the disparity is
the difference between ones and zeros per bit. The autocorrelation function $C(v)$ is
defined as

$$C(v) = \sum_{j=1}^{N} a_j a_{j+v} \text{ for } v = 0, 1, \ldots, N - 1 \tag{2.11}$$

A bit sequence will be a pseudorandom bit sequence (PRBS) if its autocorrelation
function is (Lesecq and Barraud 2008)

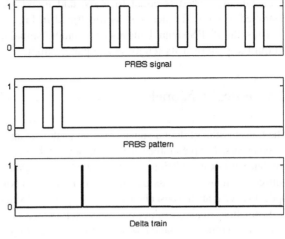

Fig. 2.9 PRBS as a composition of its pattern 0110100 and delta train

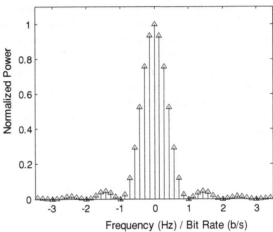

Fig. 2.10 NRZ frequency spectrum for PRBS with pattern length of 7 bits

$$C(v) = \begin{cases} m, & \text{if } v = 0 \\ mc, & \text{if } v = 1, 2, \ldots, N - 1 \end{cases} \qquad (2.12)$$

That is, the autocorrelation function shows only two possible values: m when all ones are added and mc otherwise. It means that there is no correlation among the bits of the sequence, as observed in a real random sequence. The sequence is pseudorandom, because its random pattern is limited to N bits, and from there on, the sequence starts to repeat itself, as shown in Fig. 2.9.

In the previous section, the frequency spectrum of a random binary signal was explained. However, the PRBS signal has a repetitive pattern, and hence, the non-periodicity of the signal is not satisfied. Thus, the PRBS spectrum is the combination of the pattern spectrum, which behaves as a random NRZ signal and a delta train with a period equal to $N\,T_b$. The delta train transform is also a delta train, but the period is equal to R_b/N, and therefore, the sinc^2 function is now restricted to some frequencies, as illustrated in Fig. 2.10.

Thus, the PRBS signal behaves as a random signal due to the lack of correlation within the pattern if its length is adequately high.

2.2 Optical Channels

The transmission of digital signals over an analog channel has been widely studied for several years. The preferable carrier to achieve high speed and long haul is the light, as it combines a fast modulation capability due to a short wavelength and a low interaction with certain materials (glass, plastic, air, etc.), which leads to a low attenuation. Therefore, optical systems are the best choice for high-speed and long-distance digital transmission.

From historical introduction (Sect. 1.1.1), we could already conclude that a physical channel is mandatory to achieve very high speed operation and avoid interferences due to weather conditions or opaque objects. Nevertheless, the air is usually employed as an optical channel to provide mobility where high speed and interferences are not indispensable. Remote controls are a good example of such an application. Infrared signals are usually employed because they are not visible for human eye and owing to its complete immunity for living beings.

Thus, the development of a physical optical channel—optical fiber—is a big challenge to provide a medium to transmit high-speed digital signals. The general fundamentals and the main classes of optical fibers are presented in the following section.

2.2.1 Fundamentals of Optical Fibers

Optical fiber is a cylindrical transmission channel basically consisting of a core and a cladding. Hence, a dielectric waveguide is formed connecting the emitter and receiver (Kawai 2005). Independent of the material that is made of and the selected wavelength, transmission of light along the optical fiber is based on the same principle, namely, the total reflection effect, which is a particular case of the refraction phenomenon.

The refractive index n is defined as the ratio between the speed of light through the vacuum c and the considered material v.

$$n = \frac{c}{v} = \sqrt{\frac{\varepsilon\mu}{\varepsilon_0\mu_0}} \tag{2.13}$$

It directly depends on two parameters, namely, dielectric constant ε and magnetic permeability μ, which are constant if a linear, homogeneous, and isotropic medium is assumed.

Fig. 2.11 **a** Reflected and refracted beam and **b** total reflection effect

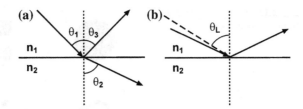

Fig. 2.12 Light transmission in optical fiber

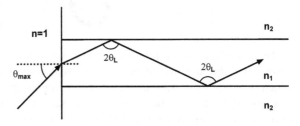

Thus, when an incident beam reaches the border between two different refractive indices, reflection and Snell's laws govern the process:

$$\theta_1 = \theta_3 \tag{2.14}$$

$$n_1 \sin \theta_1 = n_2 \sin \theta_2 \tag{2.15}$$

As we can see in Fig. 2.11a, the incident beam is split into two parts, namely, reflected and refracted. In particular, it must be remarked that if the beam crosses from high (n_1) to low (n_2) refractive index, the refracted beam (θ_2) is closer to the border than the incident one (θ_1). Therefore, there is a limit angle (θ_L) defined by

$$\theta_2 = 90° \Rightarrow \theta_1 = \theta_L = \arcsin \frac{n_2}{n_1} \tag{2.16}$$

For incident angles higher than the limit angle, refraction is not possible, thus, the beam is completely reflected. This process is continuously repeated along the optical fiber, as shown in Fig. 2.12, because the incident angle is invariant due to the cylindrical geometry.

Therefore, optical fiber propagates the light within a certain cone, denominated as acceptance cone and defined by the acceptance angle (θ_{max}). The numerical aperture, which is a dimensionless parameter that characterizes the accepted range of angles, can be calculated from the refractive indices of core and cladding.

$$N_A = n \sin \theta_{max} = \sqrt{n_1^2 - n_2^2} \tag{2.17}$$

Unlike the solution for the optical wave into an infinite medium (Wangsness 1986), the propagated signal through the fiber can be written as

$$\Psi(\vec{r}, t) = \Psi_0(x, y) e^{i(\omega t - k_g z)} \tag{2.18}$$

Fig. 2.13 **a** Single mode and
b multi mode fiber

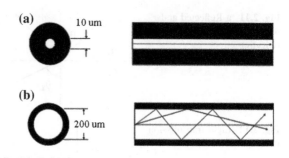

There are three main differences when we compared optical wave with a free
wave. First, the direction of propagation, that is the fiber, has been defined as z,
which is one of the Cartesian axes. Second, due to internal reflection, a stationary
wave is established in the perpendicular plane to z. Thus, particular amplitude
distributions Ψ_0 are formed according to the following condition: No transmitted
wave out of the cladding. Finally, k_g is the projection of the k-vector to the z-axis,
which is the only direction where the signal is transmitted.

Ideally, the optical fiber transmits the light with no attenuation, as the refractive
index is assumed as pure real. Nevertheless, there are undesirable effects that cause
an imaginary component of refractive index and, thus, an exponential attenuation
of the transmitted power (Wangsness 1986).

Another non-ideality is the dispersion of the fiber (Kawai 2005). In fact, dispersion can be a more restrictive parameter for the length of the fiber, as a length–
bandwidth product is derived from the total dispersion. We will now study the
three main kinds of dispersion.

Chromatic dispersion is caused by a dependency between the refractive index
and the wavelength. As light is not purely monochromatic, transmitted pulses are
progressively dispersed due to a slight variation in the transmission speed v (2.13).
A similar effect exists on transmitted pulses through optical fibers, as the value of
k_g (2.18) depends on the carrier frequency. If several modes are propagated, this
effect is much worse, as the value of k_g depends on each mode. In this case, it is
denominated as modal dispersion.

The last type of dispersion is denominated as polarization-mode dispersion
(PMD). It is caused by a defect during manufacturing. The cross-section of optical
fibers inevitably suffers from slight asymmetries, appearing as an ellipse rather
than a circle. This effect is exacerbated if the fiber senses asymmetric stresses after
installation. The resulting asymmetry yields different propagation velocities for
different polarization modes, causing the PMD.

With these effects in mind, let us explore the theoretical classes of fibers.

Fig. 2.14 a Step index and
b graded index fiber

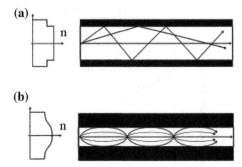

2.2.1.1 Single Mode vs. Multi Mode Optical Fiber

As mentioned earlier, *particular amplitude distributions Ψ_0 are formed* (2.18). The
number of amplitude distributions, denominated as modes, is the difference
between single and multi mode fibers, as illustrated in Fig. 2.13.

The number of modes depends on the core diameter. For adequately small core
(10 μm), only one mode is possible, that is, single mode fiber. If not, several modes (at least
two) travel along the fiber. Each mode shows a different k-vector, and hence, its projection
changes. This is represented in Fig. 2.13 as two beams traveling by different ways.
Therefore, although the modes are synchronized at the beginning, due to its different
transmission speed along the fiber, they do not arrive at the same time. In conclusion, in
contrast to single mode fibers, multi mode fibers suffer from modal dispersion.

2.2.1.2 Step Index vs. Graded Index Optical Fiber

To minimize the drawback of multi mode fibers, explained as modal dispersion
earlier, a new kind of fiber is introduced, namely, the graded index fiber. Till now, the
fiber was considered to be formed by two different refractive indices; in other words,
the fiber was considered as step index fiber. In graded index fiber, the refractive index
changes continuously from the center to the cladding, as shown in Fig. 2.14.

Thus, although several modes are excited in the fiber, the graded refractive
index is designed to compensate the different traveling distance of each mode,
minimizing the modal dispersion. Equalization can be another alternative to
minimize this dispersion (Wu et al. 2003).

In addition to the classes of fiber regarding the refractive index distribution, the
main difference between commercial fibers is the material with which it is made
of: Glass or plastic. This difference changes everything (Ziemann et al. 2008).
First, the fabrication cost is much higher for glass optical fiber. In addition,
optoelectronic connectors are simpler for plastic optical fiber, which facilitates the
installation or even lets the user to install himself/herself. However, plastic optical
fiber (POF) is not suitable for long-haul communications due to its attenuation and

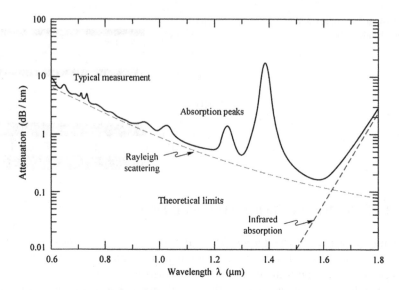

Fig. 2.15 Attenuation along a typical single-mode glass optical fiber

length-bandwidth dependency. Therefore, the choice depends basically on the application. Let us now discuss about each of these fibers.

2.2.2 Glass Optical Fibers

Glass optical fibers (GOF) are typically formed by silica (SiO_2), although doping materials, such as germanium, aluminum, or boron, are induced to modify its refractive index.

In GOF, the minimum attenuation is dominated by two opposite effects; Rayleigh scattering decreases depending on the wavelength, while infrared absorption increases. The optimum region is located from visible range to near-infrared. In addition, absorption peaks may be caused by impurities, especially OH^- ions. Fortunately, fabrication processes achieve a really low attenuation, very near to the theoretical one, as illustrated in Fig. 2.15.

Historically, the choice of the wavelength for long-haul transmissions was made by minimizing the attenuation along the optical fiber, wherever photodetectors and lasers were available. Therefore, three main windows ($\lambda = 850$, 1,300, and 1,550 nm) were consecutively employed, as presented in Table 1.1.

An attenuation as low as the theoretical limit is only required for long-haul transmission. For such an application, dispersion must also be minimized to attain high data rate. Thus, single-mode fibers operating at 1.55 μm window are preferable. If the requirements of distance and speed are less demanding, then multimode fiber may fulfill them at a lower cost. In addition, operation at visible range

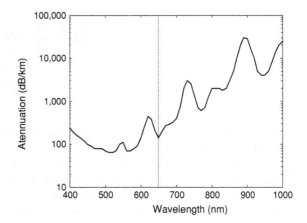

Fig. 2.16 Attenuation along a typical PMMA plastic optical fiber

(0.85 μm window) further reduces the cost of electronic devices. For short-reach applications, POF becomes a viable option.

2.2.3 Plastic Optical Fibers

The first POF were manufactured as early as the late 1960 s, when the GOF were developed. However, long-haul applications were dominated by GOF due to the lower attenuation while the requirements of short-reach transmission were fulfilled by cooper cables. There was hardly any demand for an optical medium for high data rates and small distances so that the development of the POF was slowed down for many years. Nowadays, data transmission at multi-gigabit rate over a few meters is demanded, making POF interesting.

The material most frequently employed for POF is poly-methyl methacrylate (PMMA). It is an organic compound based on a polymer chain. POF are multi-mode fiber due to the large diameter (1 mm) of the core, and are commonly manufactured with step-index profile. A typical attenuation plot is shown in Fig. 2.16.

In contrast to GOF, the attenuation of POF usually increases depending on the wavelength from visible range. Thus, transmission at a short wavelength as low as 650 nm is preferable. This wavelength is compatible with Si photodiodes and VCSEL laser or LED. POF is also suitable for blue-laser wavelength (405 nm), which is the main reason for the increase in the capacity of the last generation of optical storage systems (HD-DVD and Blu-ray).

In short-reach communications, standard step-index plastic optical fiber (SI-POF) offers certain advantages (Ziemann et al. 2008) over copper cabling: (1) Total immunity to EMI; (2) possibility of being deployed in power-line ducts; and (3) thinner cables. Furthermore, when compared with GOF, standard SI-POF

Fig. 2.17 Basic serializer
block diagram

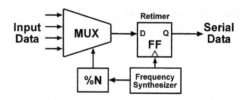

presents: (1) much simpler optoelectronic connections, (2) possibility of RCLED-
or VCSEL-based light emitters, and (3) mainly a reduction in the overall cost.

A clear disadvantage of POF is higher temperature dependence than GOF. The performances of POF are considerably degraded at temperatures higher than 85 °C.
Therefore, this issue is not critical for general optical communications as light transmission does not dissipate heat. However, some particular applications, as in automobiles,
might be affected by this degradation because the operating temperature of engines is
higher than 85 °C. POF for high-temperature conditions are under research.

Recently, 1 mm SI-POF has been standardized as A4a.2 and is beginning to be
used massively in the automotive sector (25 Mb/s) and industrial automation
(100 Mb/s) (Ziemann et al. 2008). Also, home networking is commencing; while
the required bit rate in FTTH applications is 100 Mb/s over 50 m, the goal set by
some telecom operators for home networking is to achieve speeds of several Gb/s.

2.3 Transceiver Front-End

In this section, a more detailed analysis of each block of a complete transceiver
front-end is provided: First, the transmitter, formed by the serializer and the laser
diode, and then the receiver that consists of a photodiode, receiver front-end, and
deserializer. A special attention is paid to the characteristics that affect the design
of the receiver front-end.

2.3.1 Serializer

Digital data is usually processed in packets, denominated as digital words. In
addition, several users may share the same communication link. Thus, the first
process to be carried out by the serial communication link is the combination of
low-speed parallel signals from several users in a high-speed serial signal, which
will be transmitted eventually. This is carried out by the serializer, basically
formed by a multiplexer (MUX), a retime (FF), a frequency synthesizer, and a
frequency divider (%N), as shown in Fig. 2.17.

High-speed MUXs present many design challenges as they determine the
quality of the serial data delivered to the fiber (Razavi 2003). An FF-bistable

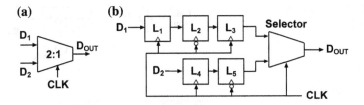

Fig. 2.18 2:1 MUX: **a** symbol and **b** typical architecture

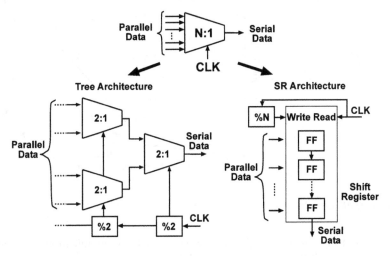

Fig. 2.19 Universal MUX: symbol (*top*) and its implementation based on tree (*left*) and shift register architecture (*right*)

before the laser driver to retime the signal avoiding jitter and other non-idealities is mandatory. As the circuit demands certain timing relationships among data input and generated clock to work properly, the design of the frequency synthesizer based on a PLL and the frequency divider (%N) must also be made carefully.

Typical 2:1 MUX architecture, shown in Fig. 2.18, includes five latchs to retime the signal before multiplexing. Thus, glitches at the output signal are avoided. Most of the MUXs are based on two topologies: A natural extension of 2:1 MUX (tree) and a direct parallel-to-serial conversion (shift-register), as shown in Fig. 2.19.

The main difference between these two types is the kind of growth depending on the number of initial parallel channels. A tree topology grows exponentially, while a shift-register based one does linearly, leading to a more compact and less power-hungry structure. However, tree architecture is dominant due to more number of inputs driven by high-speed clock and higher delay introduced by the %N divider in the shift-register counterpart. In addition, scaling down the 2:1 MUXs, in terms of power, speed, and device dimensions, attenuates the exponential growth.

Fig. 2.20 Basic transmitter
block diagram

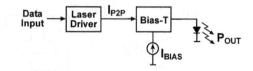

Fig. 2.21 Amplitude
modulation principle

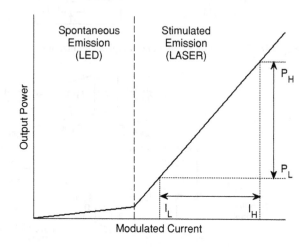

2.3.2 Laser Diode

The preferred electrical to optical converter for data communications is a laser diode (Säckinger 2005). Thus, direct modulation can be implemented, avoiding external and expensive modulator. Another cheaper light source is an LED (*Light-Emitting Diode*), but laser diodes show some advantages: mainly higher output powers and almost monochromatic and coherent light, thus, minimizing the chromatic and polarization-mode dispersion.

The operating principle of a laser is based on the stimulated emission of light. In fact, the acronym LASER means *Light Amplification by Stimulated Emission of Radiation*. The most basic structure of a laser, denoted as Fabry–Perot (FP), is formed by two parallel mirrors and an active area inside.

Thus, the laser diode must be biased to always operate where the stimulated emission is dominant against the spontaneous emission. The block diagram of the modulator circuit, shown in Fig. 2.20, is basically formed by a laser driver, which transforms the binary data signal into a two-level modulated current with a peak-to-peak value denominated I_{P2P}, a current source, which provides the biasing current I_{BIAS}, and a BIAS-T to combine both the currents.

Therefore, the transmitter is directly modulated because the current through the laser diode presents two possible values

$$I_H = I_{BIAS} + \frac{I_{P2P}}{2} \tag{2.19}$$

Fig. 2.22 Power control
system

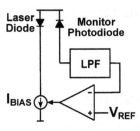

$$I_L = I_{BIAS} - \frac{I_{P2P}}{2} \qquad (2.20)$$

If the laser diode is always biased in the stimulated region, as illustrated in Fig. 2.21, where the emitted power and the modulated current are proportional, the output power P_{OUT} is also modulated according to the binary data. The coefficient χ between power and current in the laser region is called efficiency slope:

$$\chi = \frac{P_H - P_L}{I_H - I_L} \qquad (2.21)$$

It must be noted that the modulation avoids on–off switching the laser, which degrades the quality of the output power due to the transitions between stimulated and spontaneous emission, defining two positive output powers as the two possible states (High or Low).

To maintain a constant optical power in the presence of temperature variations and aging, the laser driver must employ a means of controlling the bias current. A typical power control circuit, shown in Fig. 2.22, is formed by a monitor photo-diode, a low-pass filter (LPF), and an error amplifier (Razavi 2003), which somehow controls the bias current. It must be noted that the monitor photodiode must exhibit stable characteristics with temperature and age, but its speed is not critical as it only measures the average optical power.

For long-haul applications, the FP and the distributed feedback (DFB) laser are the most commonly used lasers. DFB laser is preferable because only one wavelength is emitted, in contrast to the several wavelengths emitted by FP laser. Thus, it is ideal for wavelength-division multiplexing (WDM), although the temperature must be accurately controlled because the wavelength emitted by the DFB laser is temperature dependent. Both are considered as edge-emitting lasers because the light is emitted parallel to the wafer surface. Furthermore, 1.3 and 1.55 μm lasers can be based on InGaAsP active layer.

Short-reach applications mandate cost-efficiency of the system, and hence, non-expensive light sources are needed: Vertical-Cavity Surface-Emitting Laser (VCSEL) and LED. Both the proposals emit the light perpendicular to the wafer surface, which facilitates its integration, test, and package.

VCSEL is also a single-longitudinal mode laser similar to DFB laser. However, a wider spectral linewidth is caused by the distortion of multiple transverse modes. In addition, they are typically less powerful than DFB lasers. Owing to the high

Fig. 2.23 Optical receiver
block diagram

Fig. 2.24 Photodiode
symbol (*top*), layer structure
and band diagram (*bottom*)

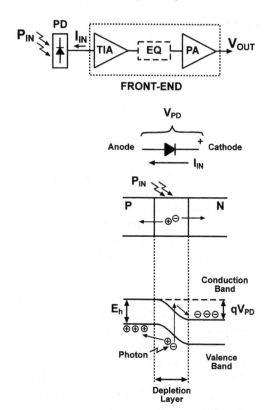

reflectivity required by the mirrors, they are commercially available at short wavelengths (up to 0.85 μm). Thus, its main application is data communications over MM-GOF or POF. VCSEL for long wavelengths is under research.

LED is based on spontaneous emission, and thus, it is not a laser. As its name indicates, the spontaneous light is emitted in all directions. Owing to the different refractive index of wafer and air, the beams emitted with a small angle to the wafer surface are confined. Thus, a vertical cone of light is formed, which is more complicated to couple to the fiber efficiently. Furthermore, the lack of a mechanism to select the wavelength leads to a much wider spectral linewidth than VCSEL. However, LED performances are superior in terms of cost and reliability than lasers. Due to its limitations, their main application is mostly in short-reach communications over MM-GOF or POF.

In recent years, resonant cavity LED (RC-LED) is under development (Bienstman and Baets 2000). It has a similar structure to VCSEL, but laser operation does not occur due to the low reflectivity of the upper mirror. The performances of RC-LED are superior to conventional LED and, although the spectrum is wider than that of a VCSEL, its temperature dependence is lower.

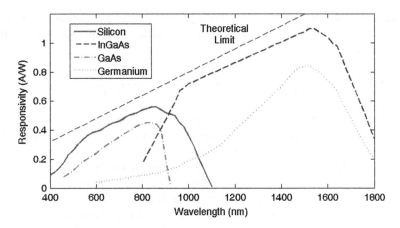

Fig. 2.25 Photodiode response for several semiconductor materials

2.3.3 Photodiode and Front-End

The power-modulated signal transmitted through the optical fiber must be converted to an electrical signal to be processed electronically. A simplified receiver block diagram, shown in Fig. 2.23, is formed by a photodiode that converts the input light into a current, and a front-end that provides an output voltage depending on the input current, formed by a transimpedance amplifier (TIA), a post-amplifier (PA), and, depending on the application, an equalizer (EQ) (Chen et al. 2007).

Photodiodes are based on the PN union (Enderlein and Horing 1997), as shown in Fig. 2.24. A reverse voltage usually is applied to the photodiode ($V_{PD} < 0$), ensuring that the device is biased in the inverse region. Therefore, the main contribution to the photocurrent comes from the electron–hole pairs generated by the incident light.

Between doped P and N layers, an area with no free carriers is formed, denominated as depletion layer. The width of this layer basically depends on the V_{PD} applied and the doping concentration, and determines the photodiode capacitance. Thus, sufficient V_{PD} or an extra insulator layer (PIN union) is usually necessary to achieve low photodiode capacitance, and hence, high speed operation.

An electron is excited to the conduction band when a photon with enough energy ($E_f = hv > E_h$) is absorbed in the union, generating a negative carrier in the conduction band and a positive carrier in the valence band. Then, both the carriers are collected by the electric field and diffusion process in the conduction band of the N layer and in the valence band of the P layer, respectively, searching the minimum energy state. Finally, the generated pair contributes to the photocurrent if an external circuit is connected between the doped layers.

The magnitude defining the photodiode performance is the responsivity R. It is quantified as the ratio between the current provided by the photodiode and the incident power

$$R = \frac{I_{IN}}{P_{IN}} = \eta \frac{\lambda q}{hc} \Rightarrow R \approx 8 \cdot 10^{-4} \eta \lambda (nm) \left[\frac{A}{W}\right] \tag{2.22}$$

As current is "charge per second" and power is "energy per second," the responsivity is also expressed as a relationship among the unity charge q, the photon energy, directly related to its wavelength λ through Plank's constant h and light speed c, and the quantum efficiency η, the probability that a photon causes a electron–hole pair. Therefore, it basically depends on the type of the photodiode, especially the material with which it is made of (Chang et al. 2005; Emsley et al. 2003), and the wavelength of the light. A comparison between the spectral responses of several kinds of photodiodes can be seen in Fig. 2.25, attaining a responsivity near to the theoretical limit ($\eta = 1$).

Figure 2.25 clearly shows that germanium and InGaAs photodiodes are suitable for 1.3 or 1.55 μm operation, while silicon and GaAs photodiodes are preferable for visible range. Silicon photodetectors integrated in CMOS chips are the most viable choice for short-reach applications due to lower cost and the advantages derived from the lack of interconnection between external photodiode and CMOS receiver; however, several drawbacks arise.

The key parameter of a semiconductor material for photonic applications is the absorption coefficient or the derived penetration depth. If the penetration depth is higher than the thickness of the absorption region leading to low quantum efficiency, a resonant cavity can be used to enhance the photodetector responsivity. Thus, optimized silicon structures attain responsivity near to the theoretical limit. However, standard CMOS technologies are not optimized for light detection. The available voltage to bias the photodiode is limited by the low supply voltage and there are no anti-reflective coatings. Furthermore, the penetration depth of light at visible range through silicon is considerably higher than the standard absorption region, which leads to a responsivity penalty.

Another disadvantage of silicon-integrated photodiodes is with regard to the speed of light detection. As some carriers are absorbed deep, where the electric field is low, they are slowly diffused, leading to low speed. In addition, these carriers cause cross-talk in an array of detectors. High-speed operation can be achieved by equalization or a differential illuminated-dark scheme. The latter is based on the fact that slowly diffused carriers will be detected by illuminated and dark diodes due to the aforementioned cross-talk, while illuminated diodes also detect high-speed carriers. Therefore, the differential signal between illuminated and dark detectors will be dominated by the high-speed carriers.

Independent of the semiconductor material, there are three contributions to the photocurrent (Säckinger 2005): I_{IN} is current that depends on the input power light, $I_{N,PD}$ is the noise contribution from the photodiode, and I_{DARK} is the constant

Fig. 2.26 Photodiode model
for inverse region

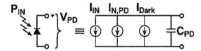

current independent of the input power light. Noise performance of the photodiode is dominated by a signal-dependent contribution

$$I_{N,PD}^2 = F \cdot G \cdot 2qI_{IN} \cdot BW_N \qquad (2.23)$$

where BW_N is the noise bandwidth associated with the measurement, which usually depends exclusively on the receiver frequency response, F is the excess noise factor, and G is a gain due to an optical preamplification or an avalanche effect.

We must remark two points about the dependency between the noise contribution from PD ($I_{N,PD}$) and the signal level (I_{IN}). First, the squared noise contribution is proportional to the signal level. In other words, if the signal level increases, the signal-to-noise ratio (SNR) improves. Second, as we explained in Sect. 2.3.2, the signal level changes between two possible positive values. Thus, noise contribution is different depending on the state, High or Low. If this asymmetry must be taken into account, the noise model becomes more complicated. Fortunately, noise from photodiode can be usually neglected (Schneider and Zimmermann 2006).

Even when the photodiode is in total darkness, a photocurrent is generated. It is denominated as dark current and depends on the junction area, temperature, and reverse voltage. This current, added to the extinction ratio (ER) of the laser, is the reason for the total photocurrent to always remain positive. We must be sure that the generated photocurrent can be correctly processed by the designed receiver. Unlike (2.23), the dark current must also be taken into account for noise contribution, but its value is usually neglected, when compared with I_{IN}.

The main parasitic impedance associated with the photodiode structure for the frequency region of interest is the depletion capacitance C_{PD}. Its value depends on the area, the reverse voltage, and the distance between the doped regions. Including all the above-mentioned components, Fig. 2.26 illustrates the considered model for photodiode biased in inverse region.

As the data signal will be processed lately as a voltage by the digital circuitry, there must be a converter added to the output of the photodiode—the front-end (García del Pozo 2010). Then, one of the main characteristics is the transresistance T_R defined as

$$T_R = \frac{V_{OUT}}{I_{IN}} \qquad (2.24)$$

In detail, the current-to-voltage conversion takes places in the transimpedance amplifier, first stage of the optical front-end. The generated voltage by the TIA is usually too small to be suitable for decision circuits, and so, a subsequent post-amplifier is required. Therefore, the transimpedance will be mainly the product of the transimpedance gain and the post-amplifier gain.

Fig. 2.27 Basic deserializer block diagram

Fig. 2.28 Typical PLL topology

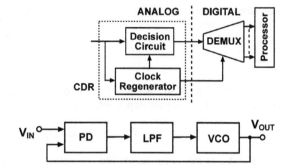

As explained in Sect. 2.1.2, random digital data shows a wide frequency spectrum. Thus, the bandwidth of the TIA is chosen to optimize the data transmission, while the post-amplifier is designed so as not to degrade the TIA frequency response in excess. Finally, a flat frequency response of the whole system is required to facilitate data recovery. If there is another cut-off frequency effect, usually due to fiber (Kawai 2005) or photodiode (Radovanovic et al. 2003), an equalizer must be included in the receiver chain to compensate such an undesirable effect.

The design of the front-end is the goal of this work. Therefore, a detailed discussion about it will be presented from the next chapter.

2.3.4 Deserializer

As shown in Fig. 2.27, deserializer is formed by a clock and data recovery circuit (CDR) and a demultiplexer.

Serial communication links do not provide a synchronization signal on a separate channel, and therefore, the receiver must rely on the extraction of the timing information from the data stream (Muller and Leblebici 2007). This clock and data recovery process can be performed in a similar way in other applications, such as electrical serial links or hard driver read-out channels. The extracted clock signal is used by decision circuit to retime the binary data signal and by subsequent circuitry for synchronization.

The decision circuitry can be as simple as a bistable circuit and the clock regenerator is usually based on a PLL due to their capability of monolithic integration. A typical PLL topology, shown in Fig. 2.28, consists of a phase detector (PD), an LPF, and a voltage-controlled oscillator (VCO) (Razavi 1996).

Other CDR topologies have also been proposed (Muller and Leblebici 2007), such as delay-locked loop (DLL), phase interpolating (PI), injection locking (IL), and gated oscillator (GO). The advantages and drawbacks of each one determine the proper topology depending on the application. The main characteristics to

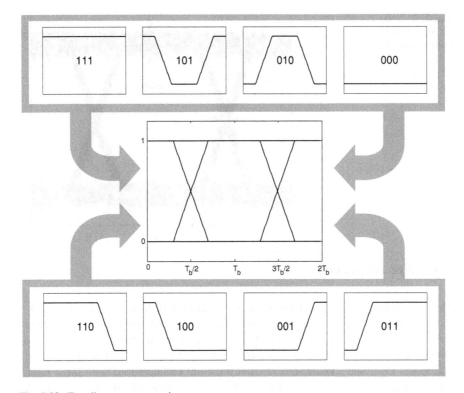

Fig. 2.29 Eye diagram construction

compare are the jitter tolerance, silicon area, power consumption and cross-talk immunity (Razavi 2002).

Finally, the serial signal is demultiplexed to process the data in parallel, according to a digital word with determinate word length and/or to divide the signal for several users.

2.4 Key Parameters

In this section, the key parameters of an optical signal transmission from receiver point of view are analyzed, namely, the bit error ratio, sensitivity, and dynamic range. They are defined in terms of input power to the receiver; however, they are determined by simulations or by experimentally measuring the signal at the output of the receiver. First, the eye diagram is presented.

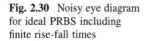

Fig. 2.30 Noisy eye diagram
for ideal PRBS including
finite rise-fall times

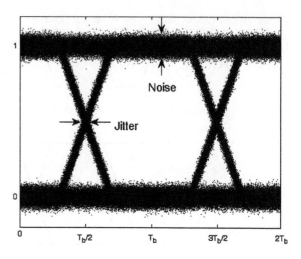

2.4.1 Eye Diagram

The eye diagram is the most common representation in time domain of a transmitted sequence. It is created, as shown in Fig. 2.29, by splitting the time twice the pulse width (T_b), to represent every bit centered. Thus, in the same plot, we are able to see many transmitted bits and all the transitions possible from the previous and to the next one. Taking into account the eight combinations with these three bits, the eye diagram for an ideal case with a certain rise-fall time is represented. Indeed, the four possibilities shown above or below the eye diagram are sufficient to represent the plot in this ideal case.

In practice, a long-enough PRBS signal is employed, including many overlapped bits in the same plot. In addition, the electrical noise is always present, degrading the eye diagram, as shown in Fig. 2.30.

Therefore, a vertical and horizontal opening can be extracted from the eye diagram, which defines the quality of the transmission. Noise is usually modeled as a Gaussian deviation from the clean value. Thus, it leads to a vertical closure of the eye diagram and random jitter, both shown in Fig. 2.30. Jitter, defined as the distribution width of cross-times at the decision level, is formed by random jitter and deterministic jitter. The latter one, caused by other undesirable effects, must be minimized to facilitate the clock and data recovery.

2.4.2 Bit Error Ratio

The bit error ratio (BER) is defined as the probability of a wrong decision at the output of the receiver. Thus, it can be expressed as

Fig. 2.31 Probability density
for Gaussian noise model

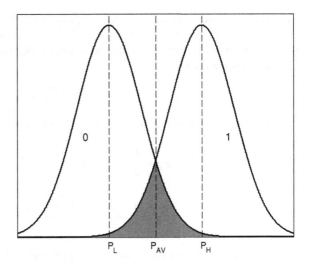

$$BER = \frac{number\ of\ erroneous\ bits}{number\ of\ transmitted\ bits} \qquad (2.25)$$

The design challenge consists of achieving a BER as low as possible, but in this attempt, two factors must be taken into account. First, if a line code is used for the transmission, errors might be detected. For example, for 6b/8b code (Widmer 2005), all 8b transmitted words consist of four "0" and four "1," and thus, the errors are detected unless a "0" and a "1" has been decided wrong in the same word, which is extremely unlikely. Then, if an error is detected due to the use of an appropriate line encoding, the improper word must be transmitted again.

Besides, a BER floor can be described, based on the life time of the receiver—whole period of time when a device should be operative—and the bit rate, according to (2.26), because a certain amount or errors can be admitted during the whole life time of the receiver. For example, for a 1 Gb/s transmission and a BER of 10^{-12}, an error is expected every 16 min of continuous transmission, and for 1 year of continuous error-free transmission, the BER should be less than 3×10^{-17}. Thus, the following relationship leads to the BER floor:

$$\frac{1}{R_b\ BER_{floor}} \ll life\ time \qquad (2.26)$$

In a real situation, the main contribution to the BER is noise, which is inherent to electronic devices. To quantify a relationship between BER and noise, a model based on Gaussian noise (Maxim Integrated Products 2008c) is described for digital transmission, where the states "1" and "0" are transmitted as pulses with

---

[3] $\int\limits_{-\infty}^{\infty} \varphi_1 dx + \int\limits_{-\infty}^{\infty} \varphi_0 dx = 1.$

P_H and P_L modulated power, respectively. For every state, a Gaussian distribution due to noise is supposed, as shown in Fig. 2.31. For simplicity, the same dispersion (σ) for the two different states and an optimum decision level [P_{AV}, defined as (2.27)] is also supposed.

$$P_{AV} = \frac{P_H + P_L}{2} \tag{2.27}$$

The probability densities of each state (φ_1 and φ_0) can be written as

$$\varphi_1 = \frac{1}{2}\frac{1}{\sigma\sqrt{2\pi}}\exp\left(\frac{-(x - P_H)^2}{2\sigma^2}\right) \tag{2.28}$$

$$\varphi_0 = \frac{1}{2}\frac{1}{\sigma\sqrt{2\pi}}\exp\left(\frac{-(x - P_L)^2}{2\sigma^2}\right) \tag{2.29}$$

The aforementioned expressions of the probability densities are normalized as if only one bit was transmitted.[3] Then, the BER is calculated as the area of φ_0 above the decision level plus the area of φ_1 below the decision level.

$$BER = \int_{P_{AV}}^{\infty} \varphi_0 dx + \int_{-\infty}^{P_{AV}} \varphi_1 dx$$

$$= \frac{1}{2}\frac{1}{\sqrt{2\pi}}\left(\int_{P_{AV}}^{\infty} \exp\left(\frac{-(x - P_L)^2}{2\sigma^2}\right)\frac{dx}{\sigma} + \int_{-\infty}^{P_{AV}} \exp\left(\frac{-(x - P_H)^2}{2\sigma^2}\right)\frac{dx}{\sigma}\right)$$

$$\tag{2.30}$$

Using these variable exchanges, we can obtain (2.31) for the first integral and (2.32) for the second one,

$$y = \frac{x - P_L}{\sigma} \Rightarrow dy = \frac{dx}{\sigma} \tag{2.31}$$

$$y = \frac{P_H - x}{\sigma} \Rightarrow dy = -\frac{dx}{\sigma} \tag{2.32}$$

For both, the following equation is satisfied:

$$x = P_{AV} = \frac{P_H + P_L}{2} \Rightarrow y = \frac{P_H - P_L}{2\sigma} = Q \tag{2.33}$$

and the Q factor can be defined as

$$Q = \frac{P_H - P_L}{2\sigma} \tag{2.34}$$

Table 2.2 Relationship between Bit Error Rate and Q factor

Q	BER	BER		BER	
		From (2.36)		From (2.37)	
0	0,5	–	–	–	–
3,090	10^{-03}	9.76E−04	2.38 %	1.09E−03	9.05 %
3,719	10^{-04}	9.87E−05	1.25 %	1.06E−04	6.45 %
4,265	10^{-05}	9.92E−06	0.81 %	1.05E−05	4.96 %
4,753	10^{-06}	9.97E−07	0.30 %	1.04E−06	4.32 %
5,199	10^{-07}	9.98E−08	0.18 %	1.04E−07	3.65 %
5,612	10^{-08}	9.97E−09	0.27 %	1.03E−08	3.00 %
5,998	**10^{-09}**	**9.97E−10**	**0.33 %**	**1.03E−09**	**2.52 %**
6,361	10^{-10}	1.00E−10	0.05 %	1.03E−10	2.59 %
6,706	10^{-11}	9.99E−12	0.12 %	1.02E−11	2.15 %
7,035	**10^{-12}**	**9.95E−13**	**0.48 %**	**1.02E−12**	**1.57 %**
7,349	10^{-13}	9.98E−14	0.25 %	1.02E−13	1.63 %
7,651	10^{-14}	9.96E−15	0.37 %	1.01E−14	1.36 %
7,942	10^{-15}	9.94E−16	0.60 %	1.01E−15	1.00 %

In bold, reference cases

Fig. 2.32 Bit Error Rate depending on Q factor

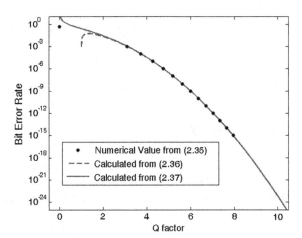

The expression of the bit error rate is reduced to only one integral, which is expected due to the symmetry of the proposed model.

$$BER = \frac{1}{2}\frac{1}{\sqrt{2\pi}}\left(\int_{Q}^{\infty}\exp\left(\frac{-y^2}{2}\right)dy - \int_{\infty}^{Q}\exp\left(\frac{-y^2}{2}\right)dy\right)$$

$$= \frac{1}{2}\frac{1}{\sqrt{2\pi}}\left(\int_{Q}^{\infty}\exp\left(\frac{-y^2}{2}\right)dy + \int_{Q}^{\infty}\exp\left(\frac{-y^2}{2}\right)dy\right) = \frac{1}{\sqrt{2\pi}}\int_{Q}^{\infty}\exp\left(\frac{-y^2}{2}\right)dy$$

$$(2.35)$$

Equation (2.35) is a non-analytical function, but it can be approximated as

$$BER \approx \frac{1}{Q\sqrt{2\pi}}\exp\left(\frac{-Q^2}{2}\right)\left(1-\frac{1}{Q^2}\right) \tag{2.36}$$

$$BER \approx \frac{1}{Q\sqrt{2\pi}}\exp\left(\frac{-Q^2}{2}\right) \tag{2.37}$$

To illustrate the dependency between Q factor and BER as well as to compare with the two analytical expressions, Table 2.2 shows the value of Q factor provided by (Säckinger 2005), which has been calculated numerically for each BER. In addition, through (2.36) and (2.37), the BER depending on the Q factor as well as the difference in percentage from the exact value of BER are calculated. It can be observed that the BER is extremely dependent on the Q factor and the analytical approximations show low error for the region of interest, that is, low BER. The data presented in Table 2.2 are visually illustrated in Fig. 2.32.

For further simplicity, the relationship between bit error rate and Q factor can be summarized using the following two equations, corresponding to the cases in bold in Table 2.2:

$$BER = 10^{-9} \Rightarrow Q \approx 6 \tag{2.38}$$

$$BER = 10^{-12} \Rightarrow Q \approx 7 \tag{2.39}$$

Furthermore, they can be easily memorized and are widely used as references.

2.4.3 Sensitivity

The sensitivity is certainly the most important parameter of the receiver (Schneider and Zimmermann 2006). It is defined as the lowest average input power for a particular BER, what indicating that it is the lower limit to properly process the transmitted signal.

$$S = P_{AV}|_{BER} \tag{2.40}$$

The definition of the sensitivity S (2.40) and the Q factor (2.34) leads to a relationship between these two parameters

$$S = P_{AV} = \frac{P_H + P_L}{2} = P_L + \frac{P_H - P_L}{2} = P_L + Q\sigma \tag{2.41}$$

Thus, the sensitivity depends on the noise performance (σ), the bit error rate through the Q factor, and the lower input power P_L. It is true if the decision level is fixed at the optimum value (P_{AV}), there is no degradation of the pulses along the receiver chain at the decision point and the Gaussian probability distribution is

---

[4] $S(dBm) = 10\log_{10}(S(mW))$.

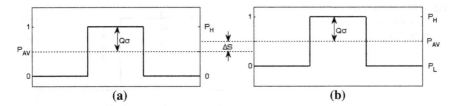

Fig. 2.33 Transmission pulse with **a** infinity and **b** non-ideal extinction ratio

Fig. 2.34 Sensitivity penalty
due to extinction ratio

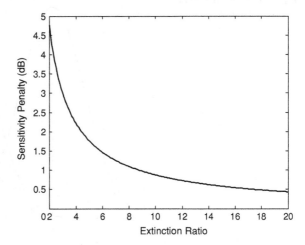

the same for both the states. Furthermore, if the probability distribution of each state is different or even is not Gaussian, then (2.41) can be extended, modifying the BER-Q relationship and defining σ as the average dispersion for both the states. The sensitivity is usually represented in dBm.[4]

This expression of sensitivity (2.41) comes from an ideal situation. We will now present the main penalty sources, namely, the ER, decision offset, inter-symbol interference (ISI), and lower cutoff frequency.

2.4.3.1 Extinction Ratio

Equation (2.41) can be modified by introducing a new parameter that is more practical, namely, the ER (Maxim Integrated Products 2008a). It is defined as the high–low power ratio

$$ER = \frac{P_H}{P_L} \tag{2.42}$$

Fig. 2.35 Asymmetry on Gaussian model due to decision offset

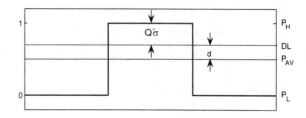

Fig. 2.36 BER contributions with decision offset

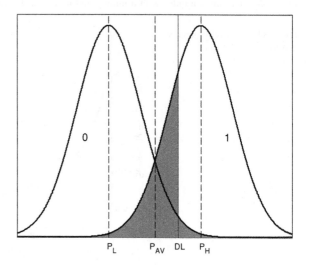

It is caused by the ac-coupling of the signal in the laser, as shown in Fig. 2.21, avoiding switching it constantly. Ideally, the ER is infinity—or at least high enough—and the sensitivity can be estimated as

$$S(ER = \infty) = Q\sigma \tag{2.43}$$

However, in reality, it is difficult to achieve an ER higher than 10. Therefore, the sensitivity is degraded by a significant factor due to ER, as shown in Fig. 2.33.

The next expression shows the relationship between the average power, the ER, and the difference of powers between each state

$$P_{AV} = \frac{P_H - P_L}{2} \frac{ER + 1}{ER - 1} \tag{2.44}$$

This expression can be rewritten to calculate the sensitivity, leading to an expression of the sensitivity penalty (ΔS), which is only dependent on the ER (2.47). Such a dependency is illustrated in Fig. 2.34.

$$S = Q\sigma \frac{ER + 1}{ER - 1} = S(ER = \infty) \frac{ER + 1}{ER - 1} \tag{2.45}$$

$$S(dBm) = S(dBm, ER = \infty) + 10 \log_{10} \frac{ER + 1}{ER - 1} \tag{2.46}$$

$$\Delta S(dB) = 10 \log_{10} \frac{ER + 1}{ER - 1} \tag{2.47}$$

In the following paragraphs, non-ideal situations are explored to attain a proper design that avoids new contributions degrading the sensitivity of the receiver and minimizing them during experimental verification.

2.4.3.2 Decision Offset

In all the previous considered cases, an optimum decision level was supposed. As shown in Figs. 2.35, 2.36, if a decision offset is present, there are two main effects: First, the sensitivity will be degraded because of the distance reduction to one of the states and second, the symmetry of the system is broken.

Thus, the BER-Q function is not valid anymore. However, fortunately, it is approximately valid between BER and Q', where

$$Q' = \frac{P_H - P_L - 2d}{2\sigma} \tag{2.48}$$

The approximation is based on the fact that one state contributes more to BER, and thus, it can be supposed that it is the only contributor. Therefore, from (2.30) with a similar exchange variable, we can obtain

$$BER \approx \int_{-\infty}^{DL} \varphi_1 dx = \frac{1}{2} \frac{1}{\sqrt{2\pi}} \int_{-\infty}^{DL} \exp\left(\frac{-(x - P_H)^2}{2\sigma^2}\right) \frac{dx}{\sigma} = \frac{1}{2} \frac{1}{\sqrt{2\pi}} \int_{Q'}^{\infty} \exp\left(\frac{-y^2}{2}\right) dy \tag{2.49}$$

When compared with (2.35), (2.50) shows a new half factor that shows that only one state contributes to BER. To observe the effect of such a factor, a very simple linear relationship between Q and BER in log scale is derived from (2.38) and (2.39) as follows:

$$\log_{10} BER = -3Q + 9 \tag{2.50}$$

From the ideal and symmetric case (2.35) to approximation (2.49), that is a half factor of difference, the increment of Q for the just-obtained linear relationship is:

$$\log_{10}\left(\frac{1}{2} BER\right) = -3(Q + \Delta Q) + 9 \tag{2.51}$$

$$\Delta Q = \frac{\log_{10} 1/2}{-3} \approx 0.1 \tag{2.52}$$

Fig. 2.37 Sensitivity penalty
due to decision offset

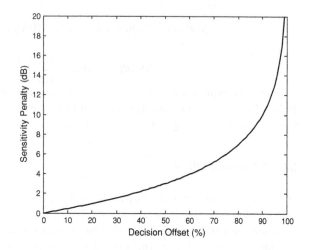

Fig. 2.38 Transmission
pulse degraded by inter-
symbol interference

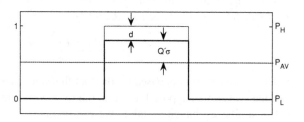

Fig. 2.39 Eye diagram
degraded by inter-symbol
interference

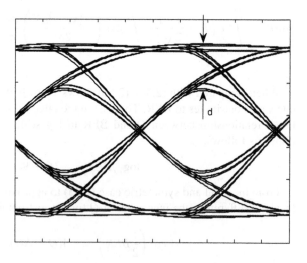

In conclusion, the same BER-Q function, where Q has been redefined, is valid
for this case with a low-enough error margin. The decision offset is normalized
according to the following expression:

Fig. 2.40 Sensitivity penalty due to inter-symbol interference

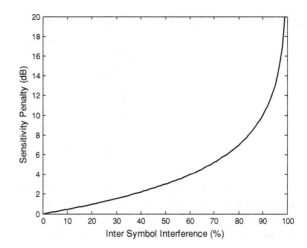

$$DL_{offset} = \frac{2d}{P_H - P_L} \qquad (2.53)$$

Now, the relationship between the sensitivity penalty and the decision offset can be derived from (2.45) with the redefined Q' factor, as follows:

$$S = Q'\sigma\frac{ER + 1}{ER - 1} \qquad (2.54)$$

As the Q factor is reduced by the decision offset (2.48), the average input power must be increased to achieve the required BER as the ER is an intrinsic property of the laser. Thus, the sensitivity is degraded by a penalty, which can be written as

$$\Delta S = \frac{Q}{Q'} = \frac{P_H - P_L}{P_H - P_L - 2d} = \frac{1}{1 - \frac{2d}{P_H - P_L}} = \frac{1}{1 - DL_{offset}} \qquad (2.55)$$

Figure 2.37 illustrates the degradation of the sensitivity due to decision offset. As we can see, for a typical worst case of 20 %, the sensitivity penalty remains below 1 dB.

2.4.3.3 Inter-Symbol Interference

Another factor that might degrade the sensitivity is the ISI. It is caused by the influence of the previous transmitted bits on the current one. It is especially important that the narrowest pulses, namely, the "010" or "101" sequences, reach the full swing. This degrading effect is illustrated in Fig. 2.38 and the ISI can be quantified as expressed in (2.56). An insufficient receiver's bandwidth is the most common cause of this effect. The derived eye diagram of such a case is illustrated in Fig. 2.39.

Fig. 2.41 Long transmission pulse degraded by lower cut-off frequency

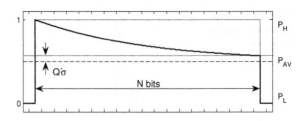

$$ISI = \frac{2d}{P_H - P_L} \tag{2.56}$$

The sensitivity penalty associated with ISI can be calculated after some approximations. It is obvious that ISI does not affect all the bits, but we can estimate that a 50 % of them are affected, because this degradation avoids reaching the full swing after a transition between different states and such a transition occurs, statistically, 50 % of the times. Therefore, as it was explained in the previous section, the Q factor is slightly modified by such a percentage affecting BER, and the relationship between these two parameters is valid, redefining Q' as

$$Q' = \frac{P_H - P_L - 2d}{2\sigma} \tag{2.57}$$

Besides, the symmetry of the system is not broken, because ISI usually degrades high–low and low–high transitions in the same way, and we considered it to be like that. Thus, the sensitivity penalty due to ISI is derived similar to the decision offset:

$$S = Q'\sigma \frac{ER + 1}{ER - 1} \tag{2.58}$$

$$\Delta S = \frac{Q}{Q'} = \frac{P_H - P_L}{P_H - P_L - 2d} = \frac{1}{1 - \frac{2d}{P_H - P_L}} = \frac{1}{1 - ISI} \tag{2.59}$$

Figure 2.40 illustrates the sensitivity penalty due to ISI given by (2.59). Owing to an analogous definition, they match with its associated ones for decision offset given by (2.55) and shown in Fig. 2.37. However, this degradation effect can be avoided by a proper design of the receiver's bandwidth, while the decision offset must be minimized during experimental verification, because it does not depend on receiver itself.

2.4.3.4 Lower Cut-off Frequency

To provide a fully differential output signal, easing the decision between "1" and "0", a control circuitry must be included in the receiver. It usually consists of a circuit to suppress the influence of background light and dark current and an offset compensation loop for the high-gain post-amplifier. Both the techniques together provide a cancellation of the input DC level owing to a lower cut-off frequency.

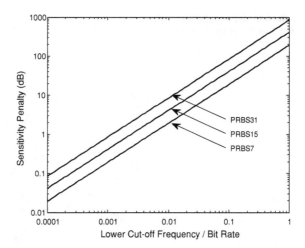

Fig. 2.42 Sensitivity penalty due to lower cut-off frequency

However, when several bits of the same state are transmitted, the sequence can be seen, for a certain period of time, as a DC signal. Such a sequence will be degraded by the lower cut-off frequency (Maxim Integrated Products 2008b), as shown in Fig. 2.41.

This undesired effect is usually minimized by scrambling with a PRBS signal or implementing a line code to the transmission as close as possible to a DC balance code. To quantify a sensitivity penalty due to this effect, a worst case is analyzed here, where a PRBS signal with no line code is considered. After N pulses of the same state, the Q factor is reduced to

$$Q' = \frac{P_H - P_L}{2\sigma} \exp\left(\frac{-N\,T_b}{\tau}\right) \tag{2.60}$$

where τ is the time constant associated with the lower cut-off frequency. Thus, for the worst case where the maximum number of bits (N_{MAX}) with the same state is present, the sensitivity shows a penalty that can be expressed as

$$\Delta S = \frac{Q}{Q'} = \exp\left(\frac{N_{MAX}T_b}{\tau}\right) = \exp\left(\frac{2\pi N_{MAX}f_{LOW}}{R_b}\right) \tag{2.61}$$

$$\Delta S(dB) = 10\log_{10}\exp\left(\frac{2\pi N_{MAX}f_{LOW}}{R_b}\right) = 20\pi\frac{N_{MAX}f_{LOW}}{R_b}\log_{10}e \tag{2.62}$$

where f_{LOW} is the lower cut-off frequency. The dependence between the sensitivity penalty in dB-scale and the lower cut-off frequency must be noted—bit rate ratio is linear, unlike the dependence due to ISI or decision offset. Such dependence is illustrated in Fig. 2.42, considering PRBS signals with different N_{MAX}.

As a rule of thumb, to limit the sensitivity penalty below 1 dB for the considered PRBS signals, the bit rate must be a thousand times higher than the lower cut-off frequency.

Fig. 2.43 Input dynamic range of the optical receiver

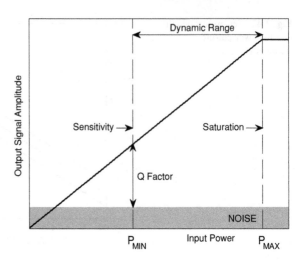

Fig. 2.44 Typical BER plot depending on the signal level

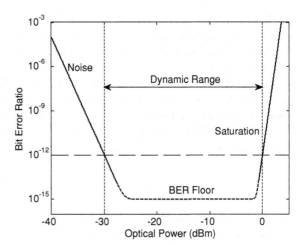

In conclusion, to determine the sensitivity:

- A proper design must be achieved, avoiding ISI and lower cut-off frequency degradation.
- The sensitivity penalty due to decision offset should be negligible.
- Then, the following three parameters are sufficient:

  - *Noise*, which can be simulated to estimate the sensitivity of the receiver (2.43) and measured to confirm the simulated noise performance.
  - *ER*, which degrades the sensitivity estimated by noise (2.47).
  - *Average input power*, to measure the sensitivity depending on the bit error rate.

Furthermore, ER and average input power are more easily measured than the high- and low-modulated powers.

Table 2.3 Reference values for optical interconnects depending on the application

Application	Long-haul	Short-reach
Wavelength	1550 nm	650 nm
Light source	FP or DFB Laser	VCSEL or LED
Optical fiber	SM-GOF	SI-POF
Attenuation	0.1 dB/km	0.14 dB/m
External photodetector	InGaAs PD	Si PD
Responsivity	1 A/W	0.5 A/W
Bit Rate	40 Gb/s	1 Gb/s
Distance	100 km	50 m

2.4.4 Dynamic Range

The sensitivity is, by definition, the lower limit for the input power, but does not provide any information about the upper limit, which is usually caused by distortion and/or saturation effects. This is because a new parameter, dynamic range (DR), must be defined (Micusik and Zimmermann 2007). It is the upper–lower limit ratio, that is, the ratio between the maximum and minimum input power properly sensed, or in other words, targeting a BER lower than a determined value. The DR is usually expressed in log scale as[5]

$$DR = \frac{P_{MAX}}{P_{MIN}} = \frac{P_{SAT}}{S} \ [dB] \qquad (2.63)$$

Thus, considering a linear relationship between the input power and the output signal level, the DR of the optical receiver can be illustrated as shown in Fig. 2.43. To experimentally validate the DR of the receiver, the BER must be measured depending on the signal level, as illustrated in Fig. 2.44, and the determined BER value leads to the lower and upper limits.

Therefore, the input DR defines the input signal level ratio, which is properly sensed. Subsequently, the more extended it is, the more independently the receiver targets the desired BER above the sensitivity level on the input signal.

2.5 Conclusions

In this chapter, the basic fundamentals of optical transmission have been introduced. They have been divided into three topics: Data format, an overview of the building blocks of the complete optical system, and the definition of key parameters.

The digital data can be formatted in many ways to be transmitted optically. NRZ is the simplest and the most used code, and hence, it is the standard to compare the state of the art. Such a code has been studied in time and frequency

---

[5] $DR(dB) = 10 \log_{10}(DR)$.

domain, and compared with RZ and 4-PAM, determining the relationship between the pulse width and its frequency response. PRBS signal is defined as the nearest generated sequence to a fully random signal.

Furthermore, all the building blocks of the optical receiver have been analyzed, such as serializer, laser diode, optical fiber, photodetector, front-end, and deserializer, although not exhaustively. The main characteristics of optoelectronic components and optical fiber, which affect receiver performance, have been introduced, such as the ER derived from the proper biasing of the LASER, the attenuation and dispersion of the optical fiber and their different kinds, the responsivity that mainly depends on the semiconductor material, and the small-signal model of the photodiode. The design and implementation of a front-end in CMOS technology is the main goal of this work, and therefore, it will be deeply explored in the following chapters.

The starting point of the last topic is the construction of the eye diagram, the most frequently used representation of the transmitted signal. The quality of the signal transmission is clearly illustrated in such a diagram, and is quantified by the BER. The relationship between noise and BER is derived, assuming an ideal transmission and a Gaussian distribution of noise. Therefore, the sensitivity of the receiver for a particular BER can be calculated from noise performance. Furthermore, some penalties derived from non-idealities are analyzed. Finally, the DR is also defined.

Thus, the contents of this chapter let us deal with the design of an optical receiver front-end, which is the focus of this book, knowing the penalties to avoid, the characteristics to optimize, and the performances to target from the reference values reported in Table 2.3, depending on the considered application.

References

Bienstman P, Baets R (2000) The RC^{2}LED: a novel resonant-cavity led design using a symmetric resonant cavity in the outcoupling reflector. IEEE J Quantum Electron 36(6):669–673

Chang S, Fang Y, Ting S, Chen S, Lin C, Lin C, Wu C (2005) Fabrication of very high quantum efficiency planar InGaAs PIN photodiodes through prebake process. IEEE Proc Circuits Devices Syst 152(6):637–640

Chen WZ, Huang SH, Wu GW, Liu CC, Huang YT, Chin CF, Chang WH, Juang YZ (2007) A 3.125 Gbps CMOS fully integrated optical receiver with adaptative analog equalizer. In: Proceedings of the 2007 IEEE asian solid-state circuits conference, pp 396–399

Couch LW (2007) Digital and analog communication systems. Prentice Hall, Upper Saddle River

Emsley M, Dosunmu O, Ünlü M, Muller P, Leblebici Y (2003) Realization of high-efficiency 10 GHz bandwidth silicon photodetector arrays for fully integrated optical data communication interfaces. In: European solid-state device research conference, 2003

Enderlein R, Horing NJM (1997) Fundamentals of semiconductor physics and devices. World Scientific, Singapore

Gantz JF et al. (2008) The diverse and exploding digital universe. International Data Corporation (IDC) via EMC, http://www.emc.com/collateral/analyst-reports/diverse-exploding-digital-universe.pdf

García del Pozo JM (2010) Design of CMOS analog front-ends for broadband optical receiver, PhD thesis, University of Zaragoza, Spain

Grigoryan V, Cho P, Godina Y, Elkridge X (2003) Novel nodulation techniques. In: Optical fiber communications conference, pp 646–647

Kawai S (2005) Handbook of optical interconnects. CRC Press—Taylor and Francis, London

Lesecq S, Barraud A (2008) A PRBS with exactly zero correlation and its application. In: IEEE 16th Mediterranean conference on control and automation, pp 724–728

Maxim Integrated Products (2008a) Extinction Ratio and Power Penalty. Application Note HFAN-2.2.0, rev. 1, 2008

Maxim Integrated Products (2008b) NRZ bandwidth—LF cutoff and baseline wander. Application Note HFAN-09.0.4, rev. 1, 2008

Maxim Integrated Products (2008c) Optical signal-to-noise ratio and the q-factor in fiber-optic communication systems. Application Note HFAN-9.0.2, rev. 1, 2008

Micusik D, Zimmermann H (2007) Transimpedance amplifier with 120 dB dynamic range. Electron Lett 43(3):159–160

Muller P, Leblebici Y (2007) CMOS multichannel single-chip receivers for multi-gigabit optical data communications, analog circuits and signal processing. Springer, Berlin

Pollard JK (1991) Multilevel data communication over optical fibre, IEEE Proceedings-I, vol 138, No 3, pp 162–168

Radovanovic S, Annema AJ, Nauta B (2003) Physical and electrical bandwidths of integrated photodiodes in standard CMOS technology. In: 2003 IEEE conference on electron devices and solid-state circuits, pp 95–98

Razavi B (1996) Design of monolithic phase-locked loops and clock recovery circuits—a tutorial. In: Razavi B (ed) Monolithic phase-locked loops and clock recovery circuits – theory and design. IEEE Press, New York, 1996

Razavi B (2002) Challenges in the design of high-speed clock and data recovery circuits. In: IEEE communications magazine, topics in circuits for communications, pp 94–101

Razavi B (2003) Design of integrated circuits for optical communications. McGraw-Hill, New York

Säckinger E (2005) Broadband circuits for optical fiber communication. Wiley, Hoboken

Schneider K, Zimmermann H (2006) Highly sensitive optical receivers. Springer Series in Advanced Microelectronics. Springer, Berlin

Shannon CE (1949) Communication in the presence of noise. In: Proceedings of the IRE, vol 37, No. 1, pp 10–21

Walker R, Dugan R (2000) 64b/66b Low-Overhead Coding Proporsal for Serial Links, http://grouper.ieee.org/groups/802/3/10G_study/public/jan00/walker_1_0100.pdf, Agilent technologies

Wangsness RK (1986) Electromagnetic fields. Wiley, New York

Widmer AX, Franaszek PA (1983) A DC-Balanced Partitioned-Block, 8B/10B Transmission Code. IBM J Res Dev 27(5):440–451

Widmer AX (1999) Partitioned DC-Balanced (0,6) 16B/18B Transmission Code, http://www.ieee802.org/3/10G_study/public/july99/widmer_2_0799.pdf, IBM

Widmer AX (2005) DC-Balanced 6B/8B transmission code with local parity, US Patent No. 6876315

Wu H, Tierno J, Pepeljugoski P, Schaub J, Gowda S, Kash J, Hajimiri A (2003) Integrated transversal equalizers in high-speed fiber-optic systems. IEEE J Solid-State Circuits 38(12): 2131–2137

Ziemann O, Kranser J, Zamzow PE, Daum W (2008) Optical short range transmission systems. In: POF handbook. Springer, Berlin

Chapter 3
Transimpedance Amplifier

The first stage of an optical receiver is usually designed as a transimpedance amplifier (TIA) since it converts the input current provided by the photodiode into an output voltage. As it is the first stage, it is the most critical component of the optical receiver, especially in terms of noise performance, as the noise produced is amplified in the latter stages. Therefore, low noise is presented as the key goal of the transimpedance amplifier in order to improve the sensitivity of the whole receiver (Schneider and Zimmermann 2006a). Other parameters, such as the bandwidth, transimpedance and total input capacitance are designed to optimize the noise performance by avoiding sensitivity penalties due to undesirable effects like inter-symbol interference.

However, the TIA does not always operate a nearby sensitive case, i.e., handling the lowest input signal considered, which is usually properly modeled by small-signal analysis. When high input signals are processed, saturation might occur and degrade the output signal significantly (Wu et al. 2004). Thus, specific techniques must be implemented in the TIA design in order to properly process as high an input photocurrent as possible, thereby enhancing the input dynamic range of the receiver.

In this chapter, theoretical fundamentals regarding the main performances of the transimpedance amplifier, such as the optimum bandwidth owing to noise—ISI trade-off, its derivation from the selected topology—shunt-feedback TIA—and the transimpedance limit is presented. A comparison with others topologies—current-mode, common-gate and regulated cascade—and an introduction to input dynamic range extension techniques is also included. Next, the proposed design implemented in a standard 0.18 μm CMOS technology suitable for low-cost applications such as POF is explained. The scalability of our proposal for CMOS technologies with shorter channel length (90 nm) is demonstrated. Finally, the verification of both prototypes is presented.

F. Aznar et al., *CMOS Receiver Front-ends for Gigabit Short-Range*
Optical Communications, Analog Circuits and Signal Processing,
DOI: 10.1007/978-1-4614-3464-1_3, © Springer Science+Business Media New York 2013

3.1 Optimum Bandwidth

Another key parameter is related with the 'speed' of the receiver, that is, the required frequency response in order to properly process the signal. The main parameter to quantify this performance is the small signal bandwidth. It must be taken into account that the transimpedance amplifier is the main noise source, since the subsequent post-amplifier will amplify it and the noise from the photo-diode is not usually dominant as mentioned in Chap. 2.

It is easy to understand that there is an optimum bandwidth owing to two opposite effects. First, too high a bandwidth leads to more noise at the output (for instance, noise from resistances that are not dependent on frequency). The contribution of this white or thermal noise is limited by the bandwidth and degrades the sensitivity of the receiver. On the other hand, if the bandwidth is too low, the input signal is filtered too much and it is not possible recover the digital signal. As previously shown in Fig. 2.7, the frequency spectrum of the NRZ signal is mostly contained within the bit rate, so the required bandwidth must be about such a value. Let us determine the optimum value more accurately.

Assuming that an input referred noise spectrum is dominated by white noise, i.e., a constant noise distribution over frequency, the total output noise is increased by a factor $\sqrt{2}$ if the bandwidth is doubled as the squared output noise is integrated over frequency. This leads to a linear dependence between sensitivity and bandwidth in the log scale with a 1.5 dB step as Fig. 3.2 illustrates. The inter-symbol interference (ISI) could add a power penalty (PP) according to (2.59). Figure 3.1c is clearly the only eye diagram affected by ISI, which can be estimated to be 50 %, leading to a 3 dB penalty. The dependence between sensitivity and bandwidth including white noise and penalty due to ISI is also illustrated in Fig. 3.2.

However, the input referred noise of a typical TIA is not only white noise. As shown in Fig. 3.3, there are three different zones depending on the shape of the power spectrum of the dominant noise contribution. At low frequencies, the dominant noise contribution is from the flicker noise showing a $1/f$ power spectrum. At high frequencies, a noise contribution with an f^2 power spectrum is expected due to a high-pass characteristic from the output to the input node within the bandwidth of the TIA, which affects the white noise contributed by the voltage amplifier (Säckinger 2005). White noise, with no frequency dependency, is dominant between these two zones.

Therefore, although the optimum bandwidth assuming white noise and inter-symbol interference caused by a Butterworth response show a particular optimum value, in practice, it is not restricted to this theoretical value due to each TIA of a particular frequency response and the input referred noise spectrum. However, it is usually limited within:

$$BW = [0.5, 0.7]R_b \qquad (3.1)$$

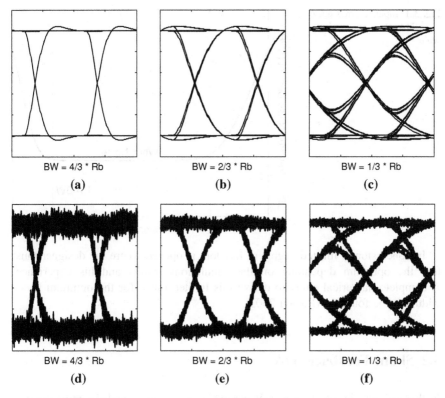

BW = 4/3 * Rb   BW = 2/3 * Rb   BW = 1/3 * Rb

(a)   (b)   (c)

BW = 4/3 * Rb   BW = 2/3 * Rb   BW = 1/3 * Rb

(d)   (e)   (f)

Fig. 3.1 Noiseless (**a–c**) and noisy (**d–f**) eye diagrams at different normalized bandwidths assuming Butterworth frequency response and white noise. **e** Represents the realistic optimum case

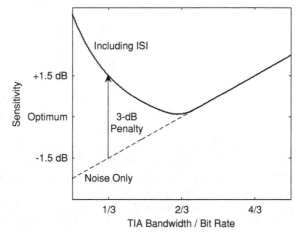

Fig. 3.2 Sensitivity depending on bandwidth including white noise and sensitivity penalty due to ISI

Fig. 3.3 Input referred noise from TIA. Both axes in log scale

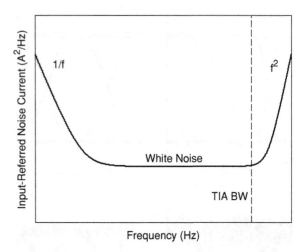

In conclusion, a limited degree of freedom is opened where the designer must find the optimum depending on the simulations results and his experience. A complete numerical study to derive this limited range for the optimum bandwidth can be found in (Razavi 2003).

3.2 Shunt Feedback TIA

In this section, the most basic I–V converter—a resistor—and the most popular structure of a transimpedance amplifier—the shunt feedback architecture—is presented and their main performances are compared (García del Pozo 2010).

First, as the target of the TIA is to convert the input current into a voltage, let us introduce the simplest I–V converter circuit: a simple resistor R_F, as shown in Fig. 3.4. The main performances—transimpedance T_R, bandwidth BW and input referred noise $I_{N,IN}$—of such a circuit can be directly derived:

$$T_R = -R_F \tag{3.2}$$

$$BW = \frac{1}{C_{PD}R_F} \tag{3.3}$$

$$I_{N,IN}^2 = I_{N,PD}^2 + I_{N,R_F}^2 \approx I_{N,R_F}^2 = \frac{4KT}{R_F} \tag{3.4}$$

where K is the Boltzmann's constant, T is the temperature and C_{PD} is the photodiode capacitance. Thus, supposing that thermal noise from the resistor $I_{N,RF}$ is dominant over noise from PD $I_{N,PD}$, the transimpedance and the noise are governed by the resistor R_F, while the bandwidth is limited by the pole associated to photodiode capacitance C_{PD} and resistor R_F. An inherent trade-off between the bandwidth and noise that is independent of R_F, as shown by the following equation:

Fig. 3.4 Simple I–V conversion by using a resistor R_F: **a** circuit and **b** equivalent model

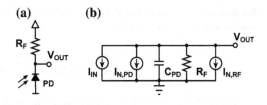

Fig. 3.5 Basic structure of a shunt feedback TIA: **a** circuit and **b** equivalent model

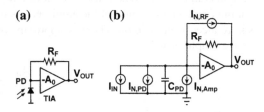

$$\frac{BW}{I_{N,IN}^2} = \frac{1}{4KTC_{PD}} = const. \tag{3.5}$$

This means that a poor noise performance is expected in order to attain high data rates and this dependence can only be minimized by reducing the capacitance of the photodiode. Therefore, to overcome this compromise between the bandwidth and noise, the TIA is introduced.

The shunt feedback TIA architecture is formed by an inverting amplifier and a feedback resistor as illustrated in Fig. 3.5. Let us analyze the same performances for such a TIA:

$$T_R = \frac{-A_0}{1+A_0} R_F \approx -R_F \ if \ A_0 \gg 1 \tag{3.6}$$

$$BW_1 = \frac{1+A_0}{C_{PD}R_F} \approx \frac{A_0}{C_{PD}R_F} \ if \ A_0 \gg 1 \tag{3.7}$$

$$I_{N,IN}^2 = I_{N,PD}^2 + I_{N,R_F}^2 + I_{N,Amp}^2 \approx I_{N,R_F}^2 + I_{N,Amp}^2 = \frac{4KT}{R_F} + I_{N,Amp}^2 \tag{3.8}$$

where the new parameters introduced by the voltage amplifier are the DC gain A_0 and the input referred noise $I_{N,Amp}$. If the DC gain A_0 of the inverting amplifier is high enough, the bandwidth[1] BW_1 of the TIA is enhanced A_0 times due to a reduction of the resistance seen from the input node, while the transimpedance does not change from the aforementioned converter. Thus, the feedback resistor R_F can be higher for the same data rate, optimizing the input referred noise in spite of

---

[1] The sub index indicates that it is calculated from a dominant-pole approximation, i.e., assuming an ideal voltage amplifier.

the new noise source, the inverting amplifier. This improvement is reflected in the bandwidth–noise trade-off:

$$\frac{BW_1}{I_{N,IN}^2} = \frac{A_0}{(4KT + I_{N,Amp}^2 R_F)C_{PD}} \tag{3.9}$$

Although the denominator increases, the trade-off is relaxed by the increase in the numerator by A_0. When the bandwidth BW_1 was calculated, it was supposed that the frequency response of the amplifier has no influence. A more realistic case is introduced from here on. The new approximation consists of a dominant pole s_a for the amplifier gain:

$$A(s) = \frac{A_0}{1 + s/s_a} \tag{3.10}$$

From (3.6) to (3.7):

$$H(s) = \frac{-R_F A(s)}{A(s) + 1 + sC_{PD}R_F} \tag{3.11}$$

a second order transfer function is derived:

$$H(s) = \frac{-R_F A_0 s_a}{(1 + A_0)s_a + s(1 + s_a C_{PD}R_F) + s^2 C_{PD}R_F} = \frac{T_R \omega_0^2}{s^2 + 2\zeta\omega_0 s + \omega_0^2} \tag{3.12}$$

characterized by a transimpedance T_R, a characteristic frequency ω_0, and a damping factor ζ given by:

$$T_R = \frac{-A_0}{1 + A_0} R_F \approx -R_F \text{ if } A_0 \gg 1 \tag{3.13}$$

$$\omega_0 = \sqrt{\frac{s_a(1 + A_0)}{C_{PD}R_F}} \approx \sqrt{\frac{A_0 s_a}{C_{PD}R_F}} \text{ if } A_0 \gg 1 \tag{3.14}$$

$$\zeta = \frac{1}{2}\frac{C_{PD}R_F s_a + 1}{\sqrt{(A_0 + 1)s_a C_{PD}R_F}} \approx \frac{1}{2}\sqrt{\frac{C_{PD}R_F s_a}{A_0}} \text{ if } A_0 \gg 1 \tag{3.15}$$

As expected, the transimpedance is not altered, but now, the bandwidth[2] BW_2 depends on the frequency ω_0 and the damping ratio ζ. As is well known, an optimum bandwidth exists for a particular value of the damping ratio:

$$\zeta = \frac{\sqrt{2}}{2} \Rightarrow s_a = \frac{2A_0}{C_{PD}R_F} = 2BW_1 \tag{3.16}$$

---

[2]   The sub index indicates that it is calculated from a second order system.

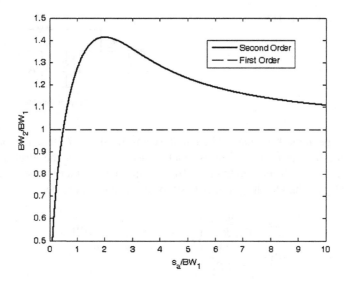

Fig. 3.6 Second order TIA bandwidth

$$\zeta = \frac{\sqrt{2}}{2} \Rightarrow BW_2 = \omega_0 = \frac{\sqrt{2}A_0}{C_{PD}R_F} = \sqrt{2}BW_1 \qquad (3.17)$$

Therefore, the bandwidth of a second-order TIA may be up to 41 % higher than that directly derived from resistor R_F and photodiode capacitance C_{PD} (3.7). In general, a biquadratic function must be solved:

$$|H(jBW_2)|^2 = \frac{T_R^2}{2} \Rightarrow BW_2^4 + \left(4\zeta^2 - 2\right)\omega_0^2 BW_2^2 - \omega_0^4 = 0 \qquad (3.18)$$

The solution can be written as

$$BW_2 = \omega_0 \sqrt{\left(1 - 2\zeta^2\right) + \sqrt{1 + \left(1 - 2\zeta^2\right)^2}} \qquad (3.19)$$

and with the approximated Eqs. (3.7), (3.14) and (3.15)

$$\frac{BW_2}{BW_1} \approx \sqrt{\frac{s_a}{BW_1}} \sqrt{\left(1 - \frac{s_a}{2BW_1}\right) + \sqrt{1 + \left(1 - \frac{s_a}{2BW_1}\right)^2}} \qquad (3.20)$$

that means that the bandwidth variation only depends on the ratio between the pole of the amplifier and the pole associated to the input node. Figure 3.6 illustrates such a dependency. It can be seen that an improvement is achieved except for the low s_a/BW_1 ratio and the optimum matches that are expected from (3.16) to (3.17).

Fig. 3.7 Current-mode TIA

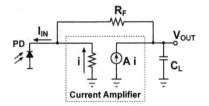

It must be remarked that, although it seems from (3.17) that the optimum bandwidth BW_2 can be increased as much as we want as for the ideal amplifier (3.7), combined with the condition (3.16), a limit appears that is associated with the limited gain–bandwidth product GBW of the voltage amplifier:

$$
\left.
\begin{aligned}
s_a &= \frac{2A_0}{C_{PD}R_F} \\
BW_2 &= \frac{\sqrt{2}A_0}{C_{PD}R_F}
\end{aligned}
\right\}
BW_2^2 = \frac{A_0 s_a}{C_{PD}R_F} = \frac{GBW}{C_{PD}R_F}
\tag{3.21}
$$

The latter equality can be rewritten as the expression known as the transimpedance limit (Säckinger 2010), that is, it is the highest transimpedance realizable depending on the technology, photodiode and bit rate:

$$
R_F \leq \frac{GBW}{C_{PD}BW_2^2}
\tag{3.22}
$$

As a conclusion, (3.22) suggests that the product of transimpedance, photodiode capacitance and the squared bandwidth is a figure of merit that is useful to fairly compare the shunt-feedback TIAs implemented in the same technology.

3.3 Review of TIA Topologies

In the previous section, the most popular TIA structure was explored and its advantages were compared to the most basic front-end, a simple resistor. However, there are some alternatives to the implementation of the TIA circuit, such as the current-mode TIA, the common-gate amplifier and the regulated cascade architecture.

Although all alternatives are based on the inclusion of current-mode circuits, the properly denominated current-mode TIA is formed by a current amplifier and a feedback resistor, as shown in Fig. 3.7. In contrast to the shunt-feedback TIA, the replacement of the voltage amplifier for a current amplifier modifies the critical node to determine the bandwidth. Ideal current amplifiers show no input and infinite output impedance and so the dominant pole is formed by feedback resistance R_F and load capacitance C_L. Therefore, this technique is preferable if the load capacitance is lower than the input capacitance (Säckinger 2010).

The primary drawback of the current-mode TIA is that it contains more noise sources than the corresponding voltage-mode TIA. In addition, voltage amplifiers

Fig. 3.8 Common-gate TIA

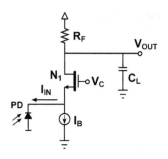

on CMOS technologies are closer to the ideal condition than current amplifiers, and this problem worsens with scalability due to short channel effects.

The second alternative is based on the only single amplifier with low input impedance, that is, the common gate (CG) structure shown in Fig. 3.8. Due to the absence of feedback loop, this TIA is known as feed-forward or open-loop TIA.

The open-loop TIA can be considered separately as a current buffer (N_1) and an I–V converter (R_F). The current buffer isolates the input capacitance from the resistor R_F and so the speed limitation of the simple I–V converter (3.3) is skipped. The transfer function of the CG stage can be modeled by two independent poles associated to the input (C_{PD}/g_m: photodiode capacitance divided by MOS transconductance) and output ($R_F C_L$) nodes.

Unfortunately, the tight trade-offs in the common gate circuits make achieving low noise difficult (Razavi 2003). Compared to the shunt feedback TIA, the feed-forward TIA is simpler and stable for any value of C_{PD}, but it tends to be noisier.

The last alternative is an improved version of the CG stage that is designed to overcome the drawback of low transconductance at low bias currents, especially in CMOS technologies. The regulated cascode circuit includes a booster amplifier A as shown in Fig. 3.9.

Thus, the effective transconductance of the MOS transistor is increased by $(A + 1)$ times of that which reduces the input capacitance in the same factor by improving the isolation of the input capacitance. Nevertheless, more noise sources are added.

In conclusion, these alternatives—and its combination with the shunt-feedback TIA—offer stability for a wide range of photodetectors, as a low input resistance suppresses the different values of photodiode capacitance. However, a noisier architecture is the price. That is why these topologies are dismissed and our proposal is based on the aforementioned shunt-feedback structure, but including input dynamic range extension. Different techniques to extend the input dynamic range are explained hereafter.

3.4 Input Dynamic Range Extension Techniques

As introduced in Sect. 2.4.4, the dynamic range is defined as the maximum to minimum signal that is properly sensed. In particular, for transimpedance amplifiers, a current ratio or power ratio can be used. The maximum current admitted by

Fig. 3.9 Regulated cascode
TIA

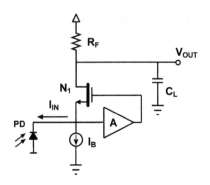

the TIA is due to overload and the minimum is directly related with the sensitivity. Thus, the dynamic range of the shunt-feedback TIA can be estimated, from the responsivity of the photodiode R and the limited output swing V_{SW}, that is the maximum voltage variation admissible at the output of the TIA, as:

$$P_{OV} \approx \frac{V_{SW}}{R \cdot R_F} \tag{3.23}$$

$$DR = \frac{P_{OV}}{S} \approx \frac{V_{SW}}{R \cdot S \cdot R_F} \tag{3.24}$$

where P_{OV} is approximately the overload input power and the sensitivity S is, by definition, related with the lowest input power that is properly sensed. Therefore, the feedback resistor directly affects the dynamic range of the TIA and the noise performance (3.8), showing an undesirable trade-off, especially for low-voltage technologies.

The DR can be extended by using a variable R_F to vary the transimpedance as a function of the input signal strength (García del Pozo et al. 2007). Alternatively, compression of the input photocurrent can be implemented, i.e., a non-linear input–output response to avoid saturation for high input signal level (Micusik and Zimmermann 2007a). A wide input dynamic range is mandatory for burst-mode receivers due to the different attenuation of different bursts and therefore they show strongly different optical power (Schneider and Zimmermann 2006b). Both methods are explained from here on. For both techniques, the considered input signal takes into account an aforementioned effect related with the laser—the extinction ratio ER.

3.4.1 Variable Feedback Resistor

To implement a variable R_F, a MOS transistor N_F working in the ohmic region (García del Pozo et al. 2007) can be used, as shown in Fig. 3.10. The transimpedance gain is reduced by the equivalent resistance of N_F:

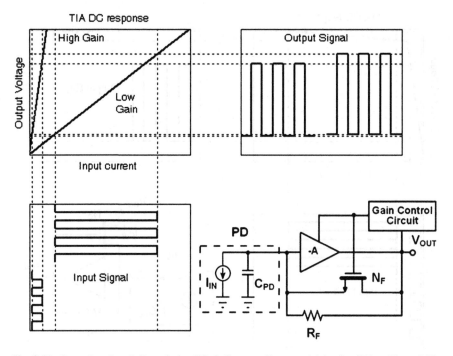

Fig. 3.10 Operation (*top-left*) and simplified diagram (*bottom-right*) of a TIA with variable feedback resistor

$$V_{OUT} = I_{IN}R_{EQ} = I_{IN}(R_F \| R_{N_F}) \tag{3.25}$$

A method to improve the control of stability and bandwidth must be introduced (Sanz et al. 2007), as both the damping factor ζ and the bandwidth are affected by the transimpedance reduction. As a typical example, if the inverting amplifier is modeled by a dominant-pole approximation, the damping factor and bandwidth depends on the transimpedance-amplifier gain ratio (Sanz et al. 2008). Thus, the amplifier gain A_0 must be also reduced to avoid instability problems.

$$A(s) = \frac{A_0}{1 + s/s_a} \Rightarrow \begin{cases} BW_2 \propto \dfrac{A_0}{R_F \| R_{N_F}} \\ \zeta \propto \sqrt{\dfrac{R_F \| R_{N_F}}{A_0}} \end{cases} \tag{3.26}$$

As shown in Fig. 3.10a, this method prevents the TIA saturation at high input currents, owing to the transimpedance reduction, but the dynamic range is usually limited due to stability affected by the variation of the damping factor. The input–output response is inherently linear with this technique. For such response, the extinction ratio only causes a sensitivity penalty (2.46).

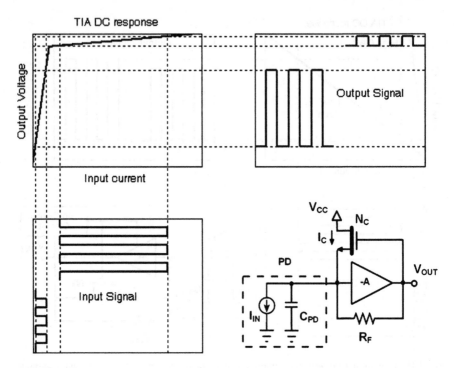

Fig. 3.11 Operation (*top-left*) and simplified diagram (*bottom-right*) of a TIA with compression technique

3.4.2 Compression Technique

Compression technique can be implemented with a MOS transistor as shown in Fig. 3.11, which introduces a current I_C depending on the output voltage. Therefore, the TIA DC response is given by (Micusik and Zimmermann 2007b):

$$V_{OUT} = (I_{IN} - I_C(V_{OUT}))R_F \tag{3.27}$$

Notice that, in spite of the denomination compression, only high input signals are compressed by a unilateral waveform clipping (Guel and Palicot 2009). For small input signals, MOS transistor N_C remains OFF and the TIA shows a linear input–output response, while higher input signals turn ON the N_C. Thus, a non-linear DC response enhances the input dynamic range. Furthermore, no additional control circuit is needed and thus this technique saves circuit complexity and power consumption. However, compression shows a problem with a large input signal and low extinction ratio: the output waveform has a large DC part and the AC amplitude is really reduced. This drawback may cause an increase of the bit error rate.

In addition to the aforementioned problem related with the low extinction ratio, the compression technique presents another drawback. Although an infinite

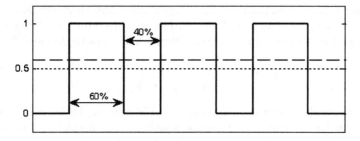

Fig. 3.12   Clock signal with duty cycle distortion due to pulse width

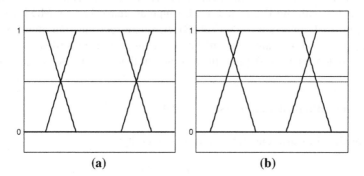

Fig. 3.13   **a** Ideal eye diagram and **b** with duty cycle distortion

extinction ratio was supposed, duty cycle distortion (DCD) would appear. Let us define this new parameter (Micusik and Zimmermann 2007a).

3.4.2.1 Duty Cycle Distortion

For a clock signal, duty cycle is defined as the ratio of the pulse duration to the pulse period. The ideal duty cycle is 50 %, but DCD might be present as illustrated in Fig. 3.12, showing a variance in timing, away from 50 % and also as an average value.

This definition can be used for pseudorandom binary sequence (PRBS) signals, but there are two differences. First, the definition is exact for signals with no disparity, while PRBS might show a disparity depending on the repetitive pattern. Since, by definition, the disparity is minimal for PRBS signals (see Appendix A.2), the definition is approximately valid for a sufficiently long pattern.

Another difference is related with rise–fall transitions. Note that, if DCD is caused by an asymmetry in the rise–fall transitions, the PRBS signal is less affected than a clock signal. The reason for this difference is that a clock signal always switches, but a PRBS sequence switches only 50 % of the times.

It was just mentioned that the average signal value depends on the DCD, that is, an average offset is caused by DCD. In fact, this distortion can be quantified, including all the causes, as the average offset—output amplitude ratio. So:

$$DCD(\%) = 100 \left| \frac{\langle V_{OUT} \rangle}{V_{OUT,MAX} - V_{OUT,MIN}} - \frac{1}{2} \right| \qquad (3.28)$$

DCD is clearly visible in an eye diagram. For instance, by including a higher pulse width for '1' than for '0', linear transitions and neglecting disparity the asymmetry reveals the DCD as illustrated in Fig. 3.13.

To explain the DCD that is inherent in the compression technique (Micusik and Zimmermann 2007a), a logarithmical model is introduced. This model is chosen because it fits with the DC response of a bipolar transistor or a MOS transistor working in weak inversion (Razavi 2008). A square root function might be more suitable for the MOS transistors in the strong inversion; nevertheless, the conclusion is independent of the non-linear function that models the compression.

3.4.2.2 Logarithmical Compression Model

The normalized logarithmical function g is given by:

$$g(x, m) = \frac{\log_a((m-1)x + 1)}{\log_a m} \qquad (3.29)$$

This normalized function (from 0 to 1 for input and output) is independent of the base of the logarithm a, so that the only degree of freedom is the factor m. It is related with the normalized transimpedance T_R for the small signal approximation, which can be derived by taking the following derivative:

$$T_R = \left. \frac{dg(x, m)}{dx} \right|_{x=0} = \frac{m-1}{\ln m} \qquad (3.30)$$

where ln means the natural logarithm. Figure 3.14 illustrates the function g for different m factors in comparison with a normalized linear function. Thus, the function g is continuous; it presents a linear region for small input signal and compresses a large input signal.

The output eye diagrams for a PRBS input sequence with a rise–fall time that represents the 40 % of the bit rate, are compared in Fig. 3.15. Due to the rise–fall transitions, the output signal for the logarithmical case presents DCD.

As previously explained, the DCD can be calculated through the average value of the output signal. Figure 3.16 shows the estimated DCD for the function g with $m = 100$ depending on the normalized input signal amplitude. According to this result, DCD is inherent to the compression technique. Therefore, a linear input–output response must be achieved to avoid DCD.

Fig. 3.14 Normalized linear
and logarithmical functions

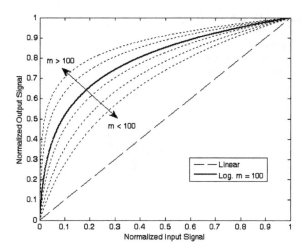

Fig. 3.15 Linear and
logarithmical eye diagrams

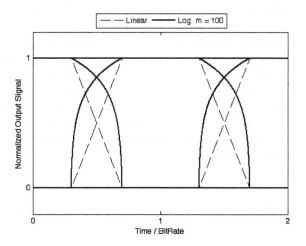

3.5 Proposed TIA Design

In this section, our proposal for the transimpedance amplifier architecture is
described. It is based on the shunt-feedback architecture, as current-mode alternatives tend to be noisier for a predetermined photodiode capacitance, as explained
previously. During the design stage, a total photodiode capacitance of 500 fF is
assumed.

In addition to the advantage of lower power consumption, low-voltage operation is necessary for current CMOS technologies. Thus, the proposed design was
implemented in two different CMOS technologies (180 and 90 nm at 1.8 and 1 V
supply voltage, respectively) to demonstrate the scalability of the most critical
building block. In practice, low-voltage operation for shunt-feedback topology

Fig. 3.16 Estimated duty
cycle distortion for
logarithmical function with
m = 100 depending on the
input signal amplitude

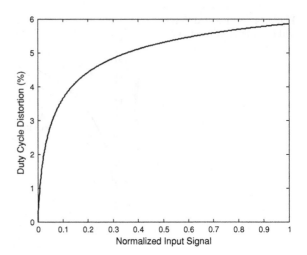

limits the possibilities of the design of the voltage amplifier and directly affects the
input dynamic range (3.24).

Finally, a modified compression technique with a linear input–output response
that improves the dynamic range with no stability problems is proposed. To protect
the TIA from saturation and improve the input current overdrive capability, the full
input photocurrent range is divided into two regions: inactive and active. In the
inactive region, the TIA responds linearly to the input current, while in the active
region the transimpedance gain can be adjusted by means of the control voltage
V_C, whereby the output voltage is still approximately linear to the input current
signal, thereby avoiding DCD.

3.5.1 180 nm Transimpedance Amplifier Architecture

The proposed TIA, shown in Fig. 3.17, is formed by a three-stage inverting
amplifier (N_1, P_1, N_2, P_2, N_3, R), a fixed shunt feedback resistor $R_F = 4.5$ kΩ, the
transistor N_4 for carrying high photocurrents and the feedback transistor (N_5).

Three stages are needed for the inverting amplifier to achieve enough gain
A with simple common source stages, which are suitable for a low-voltage operation. In this design, an inverter (N_1, P_1) is used as the first stage because it shows
the highest gain, hence optimizing the overall noise performance. The second (N_2,
P_2) and third (N_3, R) stages are common source circuits biased with a diode
connected PMOS and a resistor R, respectively. Minimal length is used in all MOS
transistors to optimize the frequency response by maximizing the transconduc-
tance-parasitic capacitances ratio. The widths of the NMOS transistors are chosen
for a good noise–power trade-off, as its transconductance depends inversely on
each other and the widths of the PMOS transistors and the value of the resistor are

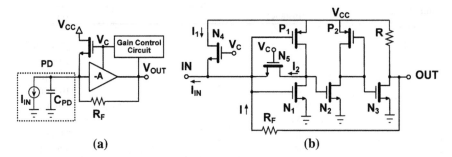

Fig. 3.17 a Simplified diagram including photodiode model and **b** complete circuit of the proposed transimpedance amplifier

designed to keep a biasing voltage of 0.9 V over the whole signal path. In particular, the list of parameters is shown in Table 3.1.

Transistor N_5 creates a current feedback path that avoids the input current overload. Transistor N_4 forms a parallel current path for high photocurrents that allows keeping a small bias current through N_1 and P_1. Both effects enhance the input dynamic range. The operation of these transistors depends on the value of their gate voltage V_C: when $V_C = 900$ mV, both transistors are OFF, in the denominated inactive region. When $V_C > 900$ mV $+ $ V$_{TH}$, both transistors are ON in the denominated active region. In this way, the TIA works in two different regions of operation depending on the control voltage V_C. A wide range of pre- and post-layout simulations demonstrates that stability is ensured over the entire input dynamic range.

3.5.1.1 TransimpedanceGain Variation

As just mentioned, the transimpedance amplifier can work in two regions: inactive and active. For $V_C = 900$ mV, both transistors N_4 and N_5 are OFF and do not affect the TIA operation in the inactive region. Therefore, the input current I_{IN} flows through the feedback resistor R_F, resulting in an output voltage given by:

$$V_{OUT} = I_{IN}R_F \quad (3.31)$$

This linear relationship, with the selected $R_F = 4.5$ kΩ and due to the limit of the output swing ($V_{SW} \approx V_{CC} - V_B \approx 0.9$ V), works properly for small input currents ($I_{IN} < 150$ μA). So, the inactive region determines the sensitivity of the transimpedance amplifier. When a higher input current is received from the photodiode, the TIA in order to avoid overload must work on the active region. This happens when N_4 and N_5 are ON, i.e., for $V_C > 900$ mV $+ $ V$_{TH}$, creating two new current paths I_1 and I_2 (see Fig. 3.17), so that:

$$I_{IN} = I_1 + I_2 + I \quad (3.32)$$

Table 3.1 Design
parameters for 0.18 μm TIA

Instance	Width/value
N_1	10 μm
P_1	27.8 μm
N_2	10 μm
P_2	27.8 μm
N_3	10 μm
N_4	5 μm
N_5	5 μm
R	719.6 Ω
R_F	4.5 kΩ

Fig. 3.18 DC response of
the TIA over the whole
control voltage swing
(inactive: $V_C = 900$ mV,
active: $V_C = 1.5$–1.8 V,
100 mV step). The input
dynamic range extension
technique causes a
transimpedance reduction
(A) and a voltage drop (B)

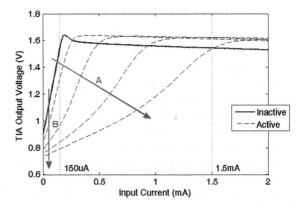

If we suppose that the voltage at the input is constant, the current I_1 is also
constant and it only depends on the voltage V_C, so we can write the input current as:

$$I_{IN} = I_1(V_C) + I_S \Rightarrow I_S = I_2 + I \qquad (3.33)$$

Then, if the transistor N_5 works in the ohmic region, we can establish the next
relationship:

$$I_2 \propto V_{DS,N5} \propto V_{OUT} \propto I \Rightarrow I_2 = \beta(V_C)I_S \qquad (3.34)$$

where $0 \leq \beta \leq 1$ is a constant that depends on the control voltage V_C, so, the
currents I_2, I_S and I are proportional to each other. Finally,

$$V_{OUT} = I \cdot R_F = [1 - \beta(V_C)][I_{IN} - I_1(V_C)]R_F \qquad (3.35)$$

This expression shows a reduction factor $1 - \beta(V_C)$ for the transimpedance gain
and a voltage drop equal to $(1 - \beta(V_C))\, R_F\, I_1(V_C)$. Both effects (denominated A and
B, respectively) that match with the simulation results as shown in Fig. 3.18.

Comparing the obtained expression (3.35) with the variable feedback resistor
(3.25) and compression (3.27) techniques, some details can be remarked. The
relationship between V_{OUT} and I_{IN} is linear for inactive and active regions thanks

Fig. 3.19 Available voltage swing depending on CMOS technology

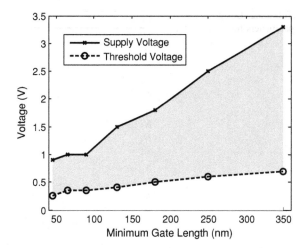

to the independent control voltage V_C. In addition, the extension of the input dynamic range is achieved thanks to a mixture of both effects: β causes a transimpedance reduction and I_1 compresses the output signal depending on the control voltage V_C instead of the instantaneous value of the output signal purely as a compression technique that would lead to a non-linear response. According to the MOS response, I_1 is a quadratic function of V_C; however, the compression effect is also affected by the β factor.

3.5.2   90 nm Transimpedance Amplifier Architecture

There is a well-known rule for deep-submicron CMOS technologies: the lower the gate length, the lower is the supply voltage (Annema 1999), thus reducing the available voltage swing. In particular, for 90 nm CMOS the supply voltage for the core is reduced to 1 V, as shown in Fig. 3.19.

Then, the width of the transistors of the proposed transimpedance amplifier must be adapted to handle such a low voltage (Aznar et al. 2009). The TIA is designed to keep a biasing voltage of 0.5 V over the whole signal path.

A lower channel length of the CMOS technology permits a higher speed operation of the circuit, leading to target a higher bit rate (Liao and Liu 2007). Instead, noise performance is improved thanks to the increase of the transconductance of the MOS transistors without increasing the parasitic capacitances. In particular, the list of parameters is shown in Table 3.2.

This supply voltage also limits the control voltage V_C. Therefore, transistors N_4 and N_5 must be designed to provide a similar input–output DC response than in the previous case with a more limited control voltage, as illustrated in Fig. 3.20.

In contrast to the 0.18 μm version, stability is not ensured over the entire dynamic range, as for $V_C = 1$ V the transient simulations show a ripple, which

Table 3.2 Design
parameters for 90 nm TIA

Instance	Width/value
N_1	30 µm
P_1	64 µm
N_2	15 µm
P_2	32 µm
N_3	15 µm
N_4	3 µm
N_5	1.5 µm
N_6	20 µm
N_7	3 µm
N_8	20 µm
I_B	500 µA
R	362.8 Ω
R_F	4.5 kΩ

Fig. 3.20 DC response of
the TIA over the whole
control voltage swing
(inactive: $V_C = 500$ mV,
active: $V_C = 700$ mV to 1 V,
100 mV step). The input
dynamic range extension
technique causes a
transimpedance reduction
(A) and a voltage drop (B)

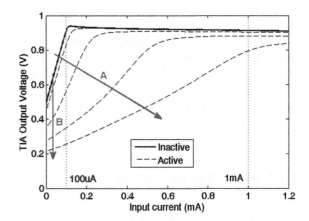

eventually dominates the behavior. In order to attain stability, a compensation
circuit (N_6, N_7, N_8 and I_B) is included (Micusik and Zimmermann 2007b). In the
inactive region, such a circuit is OFF and does not affect the TIA operation. In the
active region, as shown in the schematic in Fig. 3.21, the V_C also drives transistors
N_6 and N_7, which constitute a compensation circuit together with N_8 and I_B. This
compensation circuit helps to dynamically reduce the open-loop gain of the first
and third stages in addition to the effect of N_5 on the gain of the first stage, thus
ensuring stability over the entire input dynamic range.

3.6 Experimental Verification

In this section, the setup and results of the 90 and 180 nm transimpedance
amplifier are described. First, all the components implemented on the chip or
added to the measurement setup are explained. Then, the experimental results are

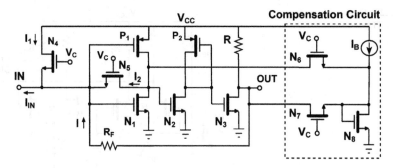

Fig. 3.21 Complete circuit of the proposed transimpedance amplifier for 90 nm CMOS technology

Fig. 3.22 Block diagram of the implemented chip for 90 nm TIA

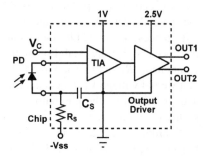

presented and discussed. The 90 nm prototype was measured only by optical characterization while the 180 nm prototype was measured optically and electrically by emulating the photodiode with an electrical circuit. Differences and advantages of both methodologies are also described and discussed.

3.6.1 90 nm TIA

The proposed transimpedance amplifier (Aznar et al. 2011) was integrated in a standard 90 nm CMOS technology with two independent supply voltages of 1 and 2.5 V for the TIA core and the output driver, srespectively. The output driver is necessary to perform experimental measurements. Its main assignment is to drive 50 Ω loads with high output swing. A differential output is highly desirable to increase supply rejection and improve noise immunity. The block diagram of the optical receiver is shown in Fig. 3.22. It includes the proposed TIA and an output driver, single-ended to differential conversion and targeting 50 Ω matching (Fig. 3.23). To create a differential output, a low-pass filter is implemented between the inputs of a differential amplifier.

Fig. 3.23 Output driver for
90 nm TIA

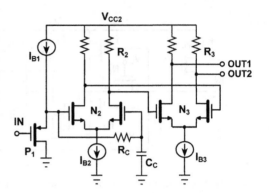

The photodiode must be biased with enough reverse voltage to attain high speed and low depletion capacitance. An external voltage might be used to properly bias the photodiode. However, the parasitic effects (resistances, capacitances and inductances) associated to both photodiode contacts might be critical to determine the speed, noise and stability of the whole receiver. Thus, the on-chip passive components R_S and C_S shown in Fig. 3.22 allow biasing the photodiode depending on external voltage V_{SS} that is not limited due to the low supply voltage of the TIA, and, furthermore, the photodiode signal is directly connected to the chip that is highly recommended.

In order to facilitate the design of the output driver with high output swing, a higher supply voltage of 2.5 V is used. Cells with this supply voltage are also available from the standard kit of 90 nm CMOS technology, which are offered for the I/O operation. A careful design might lead to implement both building blocks with 1 V, optimizing the power consumption and facilitating a higher bandwidth of the output driver, as the minimal dimensions of 2.5 V cells are higher (240 nm).

Table 3.3 summarizes the main design values for the 90 nm output driver. This circuit shows a 14 dB gain, while keeps a low-frequency cut-off below 100 kHz due to the single-ended to differential conversion.

3.6.1.1 Optical Characterization

In order to test the transimpedance amplifier optically, a key component—the photodiode—must be chosen. A low capacitance (500 fF) and high responsivity (1 A/W) was assumed, limiting the number of possibilities. Owing to such values for capacitance and responsivity, sensitivity as low as −30 dBm is attainable. In addition, a modulated light source of the selected wavelength is required.

A responsivity as high as 1 A/W can only be achieved at wavelength of 1.55 μm due to the theoretical limit (see Sect. 2.3.3). Silicon and GaAs photodiodes are not suitable for such longer wavelengths. Only InGaAs or Ge photodiodes meet the low capacitance and high responsivity requirements. A novel vertical incidence PIN Germanium photodiode (Osmond et al. 2009) was used to test the

Table 3.3 Design
parameters for 90 nm output
driver

Instance	Width/value
P_1	120 μm
N_2	216 μm
N_3	288 μm
I_{B1}	2 mA
I_{B2}	16 mA
I_{B3}	16 mA
R_2	140 Ω
R_3	50 Ω
R_C	200 kΩ
C_C	10 pF

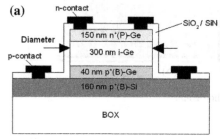
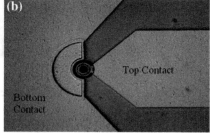

Fig. 3.24 **a** Section and **b** vertical view of Ge photodiode

designed transimpedance amplifier. A microphotograph of the photodiode is
shown in Fig. 3.24b.

This photodiode can be scaled with the diameter of the PIN Ge structure, shown
in Fig. 3.24a. The provided samples, shown in Fig. 3.25a, include photodiodes
with several diameters (25, 20, 15, 10, 7 and 5 μm). The diameter of photodiode is
directly proportional to capacitance and inversely proportional to resistance.
Therefore, a 10 μm diameter photodiode was chosen to collect the light from a
glass optical fiber with a 9 μm diameter core.

The capacitance and shunt resistance of Ge photodiodes was characterized as
shown in Fig. 3.26. Measured shunt resistance reveals that a small reverse voltage
(−0.3 to −0.1 V) is the optimum to bias the photodiode, which behaves as a
current source. For higher reverse voltage, the shunt resistance decreases dra-
matically due to a higher dark current. After calibration under open condition, the
measured photodiode capacitance for the optimum biasing is below 0.3 pF at −
0.2 V, as shown in Fig. 3.26.

The measured photodiode capacitance is lower than the supposed value during
the design stage. However, open calibration omits the parasitic capacitance
associated with metal layers, and it must be included as it is inherent to measurement. A parasitic capacitance of 0.48 pF was measured leading to a total

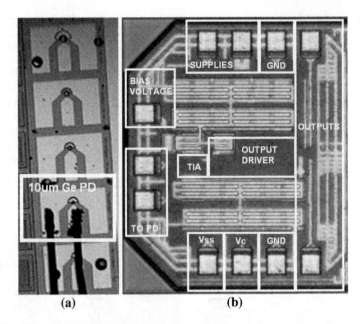

Fig. 3.25 **a** Photodiode and **b** TIA chip microphotograph

Fig. 3.26 Characterization
of Ge photodiode with 10 μm
diameter

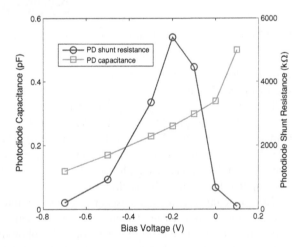

photodiode capacitance of about 0.75 pF, which is superior to the expected value.
The photodiode capacitance can be decreased with a higher reverse voltage.
Nevertheless, the aforementioned parasitic capacitance makes it difficult to target
the supposed value. Furthermore, the dark current, which should be a negligible
effect, is increased exponentially with the reverse voltage (Osmond et al. 2009).
An optimized layout to target a low parasitic capacitance is mandatory.

To avoid extra parasitic effects due to package, an on-board technique was used. This technique consists of designing a PCB where the dices are directly glued and bounded. The PCB includes capacitors to filter the input DC signals, such as supplies and bias voltages. In addition, coupling capacitors and 50 Ω matched paths to SMA connectors for outputs are added. The PCB was designed with EAGLE[3] free software, and fabricated through LeitOn Company[4] on a Rogers RO4003 substrate[5] with 0.5 mm thickness.

Some parameters can be measured to check the correct behavior of the system before introducing the input signal. By comparing with the simulation results, the DC power consumption and the output noise should match with the expected values. The measured power consumption for the TIA and driver was 4.3 and 85 mW, respectively and they match accurately with the post-layout simulation, thereby demonstrating the proper biasing of the circuit.

The measured noise performance indicates a good approximation of the simulated value. In order to estimate the noise only from the circuit, noise from other sources must be subtracted:

$$noise_{CIRCUIT} = \sqrt{noise_{ON}^2 - noise_{OFF}^2} \qquad (3.36)$$

Measurements of the RMS value with oscilloscope when both supplies ON (2.77 mV) and OFF (0.9 mV) leads to an estimation (3.36) of the noise from the circuit of 2.62 mV. This value is higher than the expected (2.47 mV) and owing to the value of the photodiode capacitance. Although a higher photodiode capacitance leads to a more reduced bandwidth, and a reduction of the noise at high frequencies might be expected, simulations demonstrate that a peak at high frequencies, which increases for higher photodiode capacitance (Säckinger 2005), degrades the noise performance.

As shown in Fig. 3.27, the high-frequency output spectral noise spectrum depends mainly on the photodiode capacitance. Then, the opposite dependency with the photodiode capacitance of bandwidth and sensitivity can be illustrated by a parametric simulation, as shown in Fig. 3.28. The sensitivity is, according to the Gaussian model introduced in Sect. 2.4.3, which is estimated from the noise performance by (3.37), where the Q-factor is equal to 7 for BER $= 10^{-12}$, T_R is the simulated midband transimpedance and a responsivity R of 1 A/W is assumed.

$$S(dBm) = 10\log_{10}\left(\frac{Q \cdot noise}{R \cdot T_R}\right) \qquad (3.37)$$

The measurement setup (Swoboda et al. 2006) includes:

---

[3]   EAGLE Cadsoft online. http://www.cadsoft.de/.

[4]   LeitOn Company. http://www.leiton.de/index.html.

[5]   Rogers Corporation 2010

Fig. 3.27 Simulated noise performance depending on photodiode capacitance

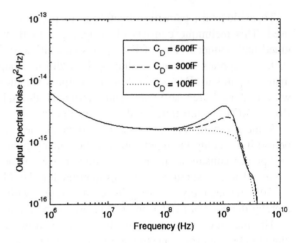

Fig. 3.28 Simulated bandwidth and sensitivity depending on photodiode capacitance

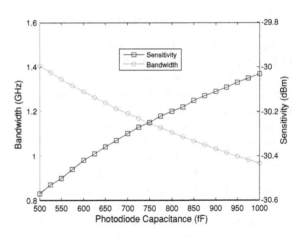

- Digital sampling oscilloscope Tektronix TDS6124C.
- Bit generator SYMPLUS RANGE BMG 12GIG.
- Clock generator CENTELLAX T G1C1-A.
- Laser ALCATEL 1915LMM.
- Laser diode controller ILX LightWave LDC3724B.
- Laser diode mount ILX LightWave Laser 2000 LDM4980.

The measured eye diagrams for several data rates are shown in Fig. 3.29. Although the transimpedance amplifier was designed to operate at 2.5 Gb/s for 500 fF, the increase of photodiode capacitance avoids reaching such a bit rate due to the bandwidth reduction shown in Fig. 3.28. These time-domain measurements demonstrate the functionality of the proposed circuit and the low noise caused by the receiver, leading to a high sensitivity.

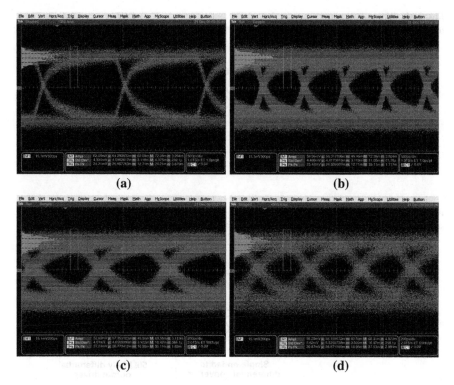

(a) (b)

(c) (d)

Fig. 3.29 Measured eye diagrams at **a** 500 Mb/s, **b** 1 Gb/s, **c** 1.5 Gb/s and **d** 2 Gb/s

3.6.2 180 nm TIA

The proposed transimpedance amplifier (Aznar et al. 2011a) was integrated in a standard 0.18 μm CMOS technology with two independent supply voltages of 1.8 V for core and output driver. Two prototypes were integrated with the only difference of the value of the shunt feedback resistor R_F − 4.5 and 1 kΩ—in order to handle a higher photodiode capacitance.

The requirements for the subsequent stages of the TIA are the same as in the previous case, that is, differential output matched to 50 Ω. The block diagram of the optical receiver is shown in Fig. 3.30. It includes the core formed by the proposed TIA with a single-ended to differential converter and a fully differential 50 Ω output driver, shown in Fig. 3.31. Table 3.4 summarizes its main design values.

The split between core and output driver is based on the different supply voltages for both prototypes, although this distinction is arbitrary as the single-ended to differential converter might be considered included in the core or in the output driver. Therefore, in this case, the TIA and the single-ended to differential converter form the core of the chip while the output driver, shown in Fig. 3.31, is a

Fig. 3.30 Block diagram of the 180 nm version

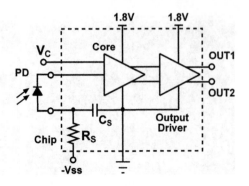

Fig. 3.31 180 nm output stage

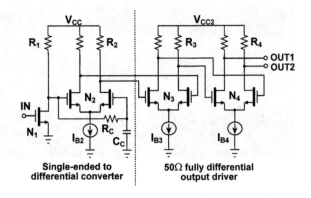

Table 3.4 Design parameters for 180 nm output driver

Instance	Width/value
N_1	10 μm
N_2	20 μm
N_3	40 μm
N_4	80 μm
I_{B2}	1 mA
I_{B3}	6 mA
I_{B4}	24 mA
R_1	450.6 Ω
R_2	1.2 kΩ
R_3	150 Ω
R_4	50 Ω
R_C	500 kΩ
C_C	5.4 pF

fully differential structure. Another remarkable difference is the intermediate stage between the TIA and the differential stages. This stage is implemented by a common-source and common-drain structure in the 180 and 90 nm versions,

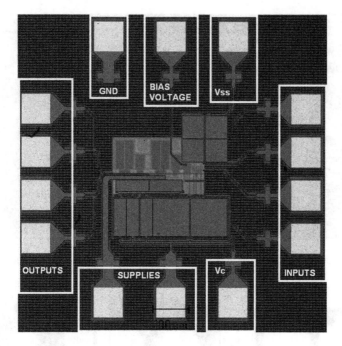

Fig. 3.32 180 nm prototype including de-embedding circuit

respectively. Thus, in each case the required level shift is achieved. Finally, the differential converter and output driver show 18.5 and 5 dB gains, respectively.

3.6.2.1 Optical Characterization

The photodiode performances are critical to test a shunt-feedback TIA, specially the photodiode capacitance. As previously shown, a higher capacitance than expected causes bandwidth shrinkage, while a lower capacitance would cause a peak at frequency response and tend to instability. Therefore, a commercial photodiode from Hamamatsu Photonics[6] was selected to measure optically the 0.18 μm integrated prototype.

In particular, S5973-01 Si PIN photodiode[7] includes a mini-lens to facilitate efficient coupling with an optical fiber. In addition, this photodiode shows high responsivity in visible range (0.44 A/W for 660 nm) while offers high speed operation (1 GHz for 50 Ω load resistance) and relatively low capacitance of 1.6 pF at 3.3 V reverse voltage. All these performances match with the required specifications to test the 0.18 μm TIA with the 1 kΩ shunt feedback resistance.

---

[6] Hamamatsu Photonics. http://www.hamamatsu.com/.

[7] Hamamatsu Photonics Si PIN Photodiode, S5971, S5972, S5973 series, Solid State Division. http://jp.hamamatsu.com/resources/products/ssd/pdf/s5971_etc_kpin1025e06.pdf.

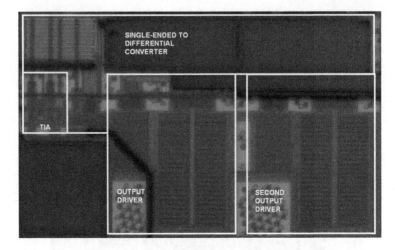

Fig. 3.33 Detail of the active area. The second output driver is added for de-embedding

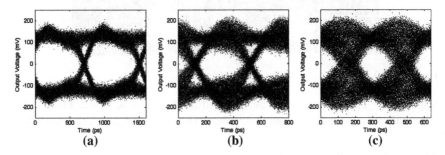

Fig. 3.34 Measured eye diagrams for PRBS $2^{31} - 1$ and an input level of -17 dBm at **a** 1.25 Gb/s, **b** 2.5 Gb/s, and **c** 3.125 Gb/s in the inactive region

The integrated prototype indicating pin-out is illustrated in Fig. 3.32, while the detail of the active area (after $180°$ rotation) is shown in Fig. 3.33. It includes a second output driver to be able to use de-embedding technique. That is the reason why there are four inputs—two for TIA and two for second driver—and four outputs.

Measurements were performed on an on-wafer probing station at room temperature and nominal conditions. The measurement setup includes:

- Digital communications analyzer Agilent 86100C.
- Bit error ratio tester Agilent N4906A.
- Commercial red laser diode from Thorlabs.[8]

The limited range of output signal amplitude provided by the sequence generator leads to a restricted variation of the modulated optical signal power. Thus, it is hard to characterize optically the dynamic range of the TIA. Nevertheless, the

---

[8] Thorlabs Inc. http://www.thorlabs.com.

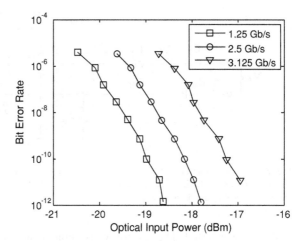

Fig. 3.35 Measured bit error rate versus input signal level

operation and the sensitivity of the TIA can be verified. The measured eye diagrams with PRBS $2^{31} - 1$ at three standard speeds (1.25, 2.5 and 3.125 Gb/s) are shown in Fig. 3.34. For all these cases, the TIA is in the inactive region as the active region is not required by the input signal level.

The measured BER versus input signal level for the three measured bit rates is shown in Fig. 3.35. Sensitivity below -17.5 dBm at 2.5 Gb/s is demonstrated indicating a BER of 10^{-12}. It must be noted that this result for sensitivity is degraded by a lower shunt feedback resistance (1 kΩ) and includes the responsivity of the photodiode (0.44 A/W).

3.6.2.2 Electrical Characterization

Optical characterization requires a more complicated measurement setup: a modulated optical source, that is, a directly modulated laser or a laser with an external modulator, a photodiode that targets the requirements and a good alignment with the optical fiber. Thus, an electrical characterization was also explored to complement the precedent optical characterization.

Unfortunately, most laboratory test signal generators do not produce current waveforms. Therefore, a method to convert a voltage waveform into a known current value was required in order to be able to test the chip without having to resort to integrating a photodiode in our prototype. A simple method is to add a large resistance in series with the input voltage signal. Characterization was done electrically for the TIA as depicted in Fig. 3.36.

The photodiode is modeled by capacitor $C_{PD} = 0.5$ pF, which corresponds to an off-chip InGaAs photodetector,[9] so the TIA including a shunt feedback resistor

---

[9] Mitsubishi Photodiodes, PD7XX7 Series. www.mitsubishielectric-mesh.com/products/pdf/pd7xx7.pdf.

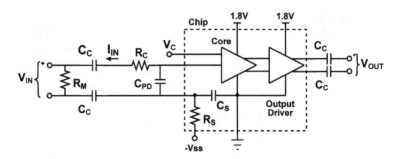

Fig. 3.36 Block diagram of the implemented chip with measurement setup for electrical characterization, including coupling capacitors C_C, an input matching resistor R_M and resistor R_C and capacitor C_{PD} modeling the photodiode

of 4.5 kΩ was tested. The resistor R_C is implemented in order to convert the input voltage into an input current without degrading the TIA performances for the inactive region. Input signal is expressed in terms of the optical power (3.38) assuming an infinite extinction ratio and a responsivity of the photodiode $R = 1$ A/ W. Coupling capacitors C_C and a 50 Ω input matching resistor R_M are also added to the measurement setup.

$$P_{IN} = \frac{I_{IN}}{R} = \frac{V_{IN}}{2R_C R} \tag{3.38}$$

The factor 2 that appears in the equation comes from the 50–50 Ω voltage divider due to input matching. As the photodiode has to be reverse biased, the biasing voltage V_{SS} is connected to the ground and the control voltage V_C is externally adjusted. Measurements were performed on an on-wafer probing station at room temperature and nominal conditions.

The core alone consumes 10.6 mW, and the total power dissipation is 64.6 mW. Frequency domain measurements were made using an R&SZVL6 vector network analyzer. The frequency response for the most critical state (inactive state determines sensitivity) is shown in Fig. 3.37. The lower cut-off frequency, as expected, is below 100 kHz, while the amplifier bandwidth with $C_{PD} = 500$ fF is about 1.55 GHz. The overall transimpedance is 67 kΩ (96.5 dBΩ).

For the time-domain measurements, a setup was used consisting of an Agilent 86100C digital communications analyzer and an Agilent N4906A bit error ratio tester. Figure 3.38 shows six eye diagrams at 2.5 Gb/s with PRBS $2^{31} - 1$ from nearby sensitivity to the highest input power considered for the adequate region: inactive region for low input power and active region for higher input powers.

As depicted in Fig. 3.38, the dynamic range of the inactive region is limited to 13 dB (from −26 to −13 dBm) due to DCD. However, the active region extends the dynamic range up to 0 dBm. In addition, eye diagrams demonstrate that DCD can be minimized thanks to the almost linear DC response of the transimpedance amplifier.

Fig. 3.37 Measured frequency response in inactive region

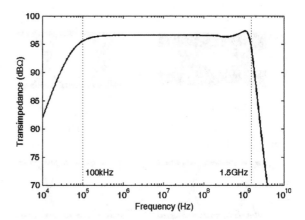

The measured BER versus input signal level for 2.5 Gb/s is shown in Fig. 3.39. These results demonstrate sensitivity below −26 dBm and optical dynamic range above 26 dB, indicating a BER of 10^{-12}.

To sum up, electrical characterization demonstrates wide input dynamic range and high sensitivity with low power consumption.

3.7 Conclusions

This chapter has covered in-depth the design of the first electrical circuit of the optical receiver: the transimpedance amplifier. The goal of this building block consists of converting the photocurrent into a voltage as efficiently as possible. Its main performances can be summarized in transimpedance, the current/voltage conversion ratio, speed, quantified by the bit rate of transmission, the equivalent input-referred noise current, leading to the sensitivity of receiver as the TIA is the main noise source, and the input dynamic range, from a lowest to highest value defined by noise performance and overload effects respectively. The bandwidth of the TIA should be from 0.5 to 0.7 times the required bit rate to achieve a noise–ISI trade-off.

As a starting point, the shunt-feedback structure, formed by a voltage amplifier and a feedback resistor, has been studied in detail, comparing its performances with the most basic I–V converter: a simple resistor. It shows a clear improvement on the bandwidth-noise ratio for the same transimpedance gain. In addition, a dominant pole associated to the voltage amplifier may enhance the bandwidth up to a 41 % for Butterworth response compared to an ideal voltage amplifier, although stability must be carefully studied as a multi-pole system is formed. Alternative TIA topologies based on current-mode techniques isolate the photodiode capacitance from the frequency performance of the TIA; however, a noisier architecture is obtained.

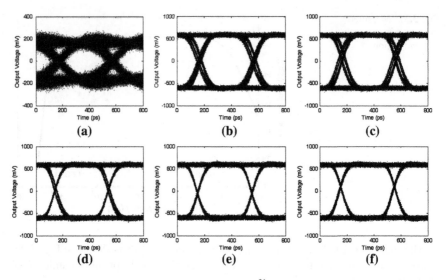

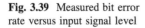

Fig. 3.38 Measured eye diagrams for 2.5 Gb/s PRBS $2^{31} -1$ for an input level of **a** -26 dBm, **b** -16 dBm, **c** -13 dBm in the inactive region and **d** -11 dBm ($V_C = 1.5$ V), **e** -1 dBm and **f** 0 dBm ($V_C = 1.8$ V) in the active region

Fig. 3.39 Measured bit error rate versus input signal level

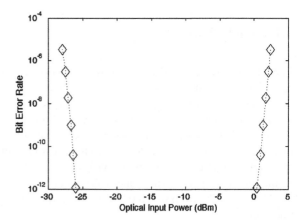

The basic shunt-feedback structure provides a limited input dynamic range. This issue can be solved by two techniques: variable feedback resistor and compression-based TIA. As a variable feedback resistor affects to the damping factor, and consequently, to the stability of the TIA, a compression technique was explored. Such a technique is associated to a non-linear input–output response that extends the input dynamic range; however, the output signal is affected by DCD and might be degraded substantially by low extinction ratio. Therefore, a new technique inspired by compression is proposed to extend the input dynamic range.

Table 3.5 Comparison of several transimpedance amplifiers

Design	Nakahara et al. (2001)	Mitran et al. (2002)	Takeshita and Nishimura (2002)	Ossier et al. (2004)	This work	
Technology	0.5 μm CMOS	0.25 μm CMOS	BiCMOS 0.6 μm/1.0 μm	SiGe 0.35 μm BiCMOS	0.18 μm CMOS	90 nm CMOS
Supply voltage	3.3 V	2 V	3.3 V	–	1.8 V	1 V
Bit rate	1 Gb/s	2.5 Gb/s	622 Mb/s	1.25 Gb/s	2.5 Gb/s	2.Gb/s[a]
Bit error rate	10^{-9}	10^{-12}	10^{-10}	10^{-10}	10^{-12}	10^{-12}
Responsivity	0.8 A/W	1 A/W	–	0.9 6A/W	1 A/W	1 A/W
Sensitivity	−28 dBm	−26.1 dBm	−29.4 dBm	−30.2 dBm	−26 dBm	−30 dBm[a]
Optical input dynamic range	18 dB	23.1 dB	29.4 dB	20.2 dB	26 dB	30 dB[b]
FOM	68.1	164.7	101.9	169.2	181.1	525
Power dissipation	<30 mW	76 mW	220 mW	–	10.6 mW	4.3 mW

[a] Partially confirmed by measurements. [b] Not measured, Simulated value

The proposed design, consisting of a three-stage inverting amplifier and a high-value feedback resistor, is suitable for low-voltage operation. Transistors' dimensions were designed to achieve a good noise–power trade-off. Two transistors driven by a control voltage were implemented to vary the transimpedance gain, according to a linear input–output response. Our proposal was integrated in two different standard CMOS technologies: 0.18 μm at 1.8 V and 90 nm at only 1 V targeting 2.5 Gb/s bit rate transmission. Superior noise performance is obtained for 90 nm at the same bit rate, but additional circuitry must be included to ensure stability over the entire input dynamic range.

Electrical and optical characterizations were carried out to test 0.18 μm and 90 nm prototypes. The selected photodiode for optical characterization of 90 nm prototype targets the depletion capacitance assumed during the TIA design. However, the large parasitic capacitance associated to photodiode layout increases the total photodiode capacitance significantly, thereby avoiding testing of the TIA properly. Measured noise performance demonstrates high sensitivity. An emulation circuit of the photodiode was proposed for the 0.18 μm prototype, attaining a full characterization of the design in the time and frequency domains. Assuming a responsivity of 1 A/W, the measurements validate a sensitivity below −26 dBm, with a slight penalty around 1 dB compared to the equivalent post-layout simulations, and a wide input dynamic range of 26 dB by properly processing the highest current provided by a typical PIN photodiode.

In order to make a fairer comparison including the sensitivity S and the dynamic range DR, a figure of merit (FOM) is defined as

$$FOM = \frac{R_B \cdot Q \cdot DR}{S} \left[\frac{Gb/s \cdot dB}{\mu W} \right] \qquad (3.39)$$

As shown in Table 3.5, the 0.18 μm TIA shows competitive results compared with previously published designs with the same data rate and bit error rate (Mitran et al. 2002) and compared with designs with lower data rates which target less demanding bit error rates (Nakahara et al. 2001; Ossier et al. 2004; Takeshita and Nishimura 2002).

Compared exclusively to works with slightly lower FOM, the photodiode capacitance of (Mitran et al. 2002) is only 275 fF and (Ossier et al. 2004) implements an avalanche of photodiodes with a gain of 6. Both effects help to improve sensitivity. Additionally, the power dissipated by the proposed TIA is remarkably lower than in (Mitran et al. 2002) while (Ossier et al. 2004) does not report the power performance.

The FOM of the 90 nm TIA is almost three times better. Nevertheless, it is calculated by some estimated results. Measured noise performance leads to an estimation of a sensitivity near to −30 dBm. A bit rate of 2.5 Gb/s is expected to be fulfilled with the appropriate photodiode capacitance. The optical input dynamic range must be confirmed by measurements. Finally, the power consumption is remarkably reduced from the 0.18 μm prototype.

References

Annema AJ (1999) Analog circuit performance and process scaling. IEEE Trans Circuits Syst II 46(6):711–725

Aznar F, Gaberl W, Zimmermann H (2009) A highly sensitive 2.5 Gb/s transimpedance amplifier in CMOS technology. In: Proceedings of the 2009 IEEE international symposium on circuits and systems, pp 189–192

Aznar F, Gaberl W, Zimmermann H (2011) A 0.18 μm CMOS transimpedance amplifier with 26 dB dynamic range at 2.5 Gb/s. Microelectron J 42:1136–1142

Aznar F, Gaberl W, Zimmermann H (2011a) A 90 nm CMOS transimpedance amplifier with − 30 dBm sensitivity at 2.5 Gb/s. IEEE Trans Circuits Syst II, 2011, Under review

García del Pozo JM, Celma S, Sanz MT, Alegre JP (2007) CMOS Tunable TIA for 1.25 Gbit/s optical Gigabit ethernet. Electron Lett 43(23):1303–1305

García del Pozo JM (2010) Design of CMOS analog front-ends for broadband optical receiver. PhD thesis, University of Zaragoza, Spain

Guel D, Palicot J (2009) Analysis and comparison of clipping techniques for ofdm peak-to-average power ratio reduction. IEEE International conference on digital signal processing, pp 1–6

Liao C, Liu S (2007) A 40 Gb/s transimpedance-AGC amplifier with 19 dB DR in 90 nm CMOS. IEEE Int Solid-State Circuits Conf 586:54–55

Micusik D, Zimmermann H (2007a) Transimpedance amplifier with 120 dB dynamic range. Electron Lett 43(3):159–160

Micusik D, Zimmermann H (2007b) A 240 MHz-BW 112 dB-DR TIA. IEEE Int Solid-State Circuits Conf 621:554–555

Mitran P, Beaudoin F, El-Gamal MN (2002) A 2.5-Gbit/s CMOS optical receiver frontend. Proc 2002 IEEE Int Symp Circuits Syst 5:441–444

Nakahara T, Tsuda H, Ishihara H, Tateno K, Amano C (2001) High-sensitivity 1 Gb/s CMOS receiver integrated with GaAs- or InGaAs-photodiode by wafer-bonding. Electronic Lett 37(12):781–782

Osmond J, Vivien L, Fédéli J-M, Marris-Morini D, Crozat P, Damlencourt J-F, Cassan E, Lecunff Y (2009) 40 Gb/s Surface-illuminated Ge-on-Si photodetectors. Appl Phys Lett 95:151116-151116-3

Ossier P, Yi YC, Bauwelinck J, Qiu XZ, Verndewege J, Gilon E (2004) DC-Coupled 1.25 Gb/s burst-mode receiver with automatic offset compensation. Electronic Lett 40(7):447–448

Razavi B (2003) Design of integrated circuits for optical communications. McGraw-Hill, New York

Razavi B (2008) Fundamentals of microelectronics. Wiley, Hoboken

Rogers Corporation (2010) RO4000® Series High Frequency Circuit Materials. http://www.rogerscorp.com/documents/726/acm/RO4000-Laminates—Data-sheet.aspx. Revised 05/2010

Säckinger E (2005) Broadband circuits for optical fiber communication. Wiley, Hoboken

Säckinger E (2010) The transimpedance limit. IEEE Trans Circuits Syst I 57(8):1848–1856

Sanz MT, García del Pozo JM, Celma S, Sarmiento A (2007) Constant-bandwidth adaptive transimpedance amplifier. Electron Lett 43(25):1451–1452

Sanz MT, García del Pozo JM, Celma S, AlegreJP, Sarmiento A (2008) Tunable transimpedance amplifiers with constant bandwidth for optical communications. Proceedings of the 2008 IEEE international symposium on circuits and systems, pp 65–68

Schneider K, Zimmermann H (2006a) Highly Sensitive Optical Receivers. Springer Series in Advanced Microelectronics

Schneider K, Zimmermann H (2006b) Three-stage burst-mode transimpedance amplifier in deep-sub-μm CMOS technology. IEEE Trans Circuits Syst I 53(7):1458–1467

Swoboda R, Schneider K, Zimmermann H (2006) Optical receivers with large-diameter photodiode. Integr Opt Silicon Photonics Photonic Integr Circuits Proc SPIE 6183:61831D

Takeshita T, Nishimura T (2002) 622 Mb/s Fully integrated optical IC with a wide range input. IEEE International Solid-State Circuits Conference, pp 258–259

Wu C, Liu C, Liu S (2004) A 2 GHz CMOS variable-gain amplifier with 50 dB linear-in-magnitude controlled gain range for 10 GBase-LX4 ethernet. IEEE International solid-state circuits conference

Chapter 4
Post-Amplifier

Following the photocurrent-to-voltage conversion performed by the transimpedance amplifier (TIA), the post-amplifier provides additional voltage gain for the signal to satisfy the input sensitivity of the clock and data recovery circuit. Owing to the signal amplification already operated in the TIA, noise is less critical at this stage (Schneider and Zimmermann 2006).

Therefore, sufficient output swing is presented as the main goal of the post-amplifier in any case. In particular, a small signal of a few millivolts is usually provided at the TIA output in the sensitive case, i.e., the lowest input signal properly processed. Thus, a high-gain post-amplifier—typically 30 dB—is required to boost signal swing to adequate level for subsequent circuitry (Razavi 2001).

Post-amplifiers for optical communications are frequently implemented with limiting amplifiers due to a simplification of the architecture, avoiding programmability and automatic gain control (AGC) loop, thus being a cost-effective choice. Furthermore, its power dissipation, bandwidth, noise, and so forth are often superior to an AGC amplifier realized with the same technology (Hermans and Steyaert 2007). However, limiting amplifiers are not always appropriate. Instead, AGC amplifiers show clear advantages due to its linear input–output response, preserving the signal waveform. Thus, analog signal processing can be performed at the output of the AGC, such as equalization, slice level steering, and soft decision decoding (Säckinger 2005). Finally, they are suitable for multi-level data transmission (Atef et al. 2008). An AGC post-amplifier is formed by three buildings blocks, namely the amplifier core, the DC compensation circuit, and the AGC loop, as shown in Fig. 4.1.

To set all the requirements, the AGC post-amplifier (Hermans and Steyaert 2007) must not degrade the bandwidth-optimized signal from the TIA and present a wide controlled gain range to provide proper operation, independent of the input signal. Thus, the bandwidth must be considerably higher over the whole dynamic range than that selected for the TIA and the gain should be able to be reduced up to 0 dB to avoid clipping of the output signal even for high input amplitude level.

F. Aznar et al., *CMOS Receiver Front-ends for Gigabit Short-Range*
Optical Communications, Analog Circuits and Signal Processing,
DOI: 10.1007/978-1-4614-3464-1_4, © Springer Science+Business Media New York 2013

Fig. 4.1 Basic post-amplifier diagram

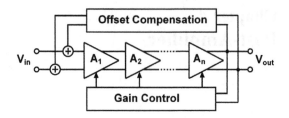

In addition, the noise caused by the post-amplifier must be taken into account; however, it usually represents only a small fraction of the total output noise for the sensitive case. Finally, as the envelope signal speed is much lower than the transmission rate, the maximum settling time of the AGC is set to 1 μs for the worst case.

In this chapter, theoretical fundamentals regarding the amplifier core, such as multistage design and broadband techniques, and auxiliary loops, such as constant settling time for AGC and offset compensation, are presented. The proposed AGC amplifier is implemented in a low-cost CMOS technology, and its design is explained step-by-step. Finally, the verification of such a circuit is included.

4.1 Amplifier Core

The two main specifications of the amplifier core are high gain and high speed, i.e., wide bandwidth. Therefore, the gain-bandwidth product, GBW, must be optimized. It must be noted that the GBW is only constant for first-order amplifiers, and hence, it becomes a good figure of merit. In this study, superior orders are included for comparison, although GBW is not strictly a good figure of merit.

Once the choice of the technology is made, there is, in principle, a boundary to design amplifiers (Razavi 2008), which is the *transition frequency* f_T of transistors, defined as the frequency where its current gain becomes unity. It is caused by the inherent parasitic impedances of the transistor and, if a first-order response is supposed, it leads to a constant product between the current gain and the bandwidth. In a similar way, the GBW of a voltage gain cell is also limited. In practice, the highest operation frequency of a transistor is only a fraction of f_T, and this parameter is only used for comparison among the technologies.

4.1.1 Multistage Structure

The requirements of the post-amplifier in terms of gain and bandwidth lead to a multistage design, relaxing the GBW of each stage (Jindal 1987). Cascading several stages, the total gain will be the product of each one while the bandwidth is not reduced in the same factor. For an ideal situation where the post-amplifier consists of n identical stages defined by a constant gain A_i within the bandwidth BW_i and no signal transmitted beyond, i.e., a brick-wall frequency response,

$$\left.\begin{array}{l} A = A_i^n \\ BW = BW_i \end{array}\right\} \quad \frac{GBW}{GBW_i} = \frac{A \cdot BW}{A_i \cdot BW_i} = A_i^{n-1} = \sqrt[n]{A^{n-1}} \tag{4.1}$$

where the parameters with no sub-index refers to the whole amplifier, while the sub-index i refers to each stage. This expression clearly shows that the gain bandwidth limitation can be skipped. Unfortunately, in a real situation, the total bandwidth is reduced with respect to the bandwidth of each stage. It can be demonstrated that for a first- and second-order Butterworth frequency response, the relationship (4.1) respectively becomes (Säckinger 2005):

$$\frac{GBW}{GBW_i} = \sqrt[n]{A^{n-1}} \cdot \sqrt{\sqrt[n]{2} - 1} \tag{4.2}$$

$$\frac{GBW}{GBW_i} = \sqrt[n]{A^{n-1}} \cdot \sqrt[4]{\sqrt[n]{2} - 1} \tag{4.3}$$

Thus, a general expression of the extension ratio of the GBW can be written, where m denotes the degree of the Butterworth frequency response:

$$\frac{GBW}{GBW_i} = \sqrt[n]{A^{n-1}} \cdot \sqrt[2m]{\sqrt[n]{2} - 1} \tag{4.4}$$

For real cases, there is an optimum number of stages n_{opt}, because the higher the number of stages is, the higher is the first term and the lower is the second term. On calculating the derivative of the general expression, we obtain:

$$n_{opt} = \left(\log_2 \left[\frac{1}{1 - \frac{\ln 2}{2m \ln A}} \right] \right)^{-1} \approx 2m \ln A \ \text{if} \ A \gg \sqrt[2m]{2} \tag{4.5}$$

From this approximated result, the optimum gain of each stage and the optimum extension ratio for the GBW can be derived as:

$$A_{i,opt} \approx \sqrt[2m]{e} \tag{4.6}$$

$$\left.\frac{GBW}{GBW_i}\right|_{opt} \approx \frac{A}{\sqrt[2m]{e}} \sqrt[2m]{\frac{\ln 2}{2m \ln A}} \tag{4.7}$$

A comparison of the four cases, namely, ideal brick wall, first-order, second-order Butterworth, and general case with $m = 5$, for a typical total gain (30 dB) of the post-amplifier, is illustrated in Fig. 4.2. The location of the optimum number of stages and the optimum extension ratio from $m = 1$ to 5 is also shown by x-marks.

It can be seen that the optimum number of stages provides a huge extension of the GBW. However, the optimum number is quite high. For such solution, the

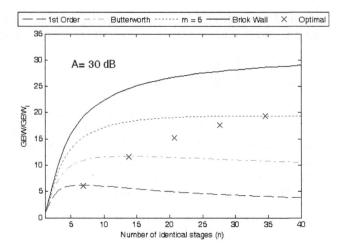

Fig. 4.2 Gain–bandwidth product extension

required bandwidth for each stage and, particularly, the power and area consumption become a serious drawback (Hermans and Steyaert 2007). Hence, the selected number of stages remains lower than the optimum, being normally restricted to 3–5 stages with gains of 6–12 dB, while to further enhance the GBW, broadband techniques are applied.

4.1.2 Broadband Techniques

A multistage structure is able to provide an optimization of the GBW, but it is caused by an increment in the total gain superior to the shrinkage of the bandwidth. Thus, some broadband techniques (Hermans and Steyaert 2005) must be applied at the gain stages to attain the bandwidth requirements. The implemented techniques, such as inter-stage buffering, inverse scaling, negative capacitances, and zero-pole cancellation, are briefly discussed.

4.1.2.1 Inter-stage Buffering

As the post-amplifier is formed by several consecutive stages, one of the contributions to the limitation of the total bandwidth can be derived from the equivalent inter-stage capacitance. Figure 4.3a illustrates this fact for two identical cascaded stages, neglecting the interconnect capacitance.

As can be seen in Fig. 4.3b, inclusion of a buffer with an input capacitance $C_B < C_I$ reduces the capacitance associated with the first node. Thus, if the buffer can be considered as ideal, i.e., it drives the second gain stage offering no frequency limitation; then the bandwidth of the inter-stage node is enhanced according to the following expression (Säckinger 2005):

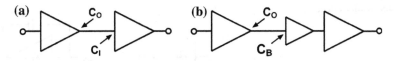

Fig. 4.3 Two identical stages **a** without and **b** with inter-stage buffer

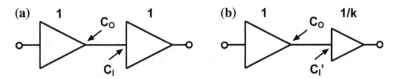

Fig. 4.4 Two consecutive stages **a** without and **b** with inverse scaling

$$BW' = \frac{C_I + C_O}{C_B + C_O} BW \tag{4.8}$$

In practice, the buffer shrinks a portion of the gained bandwidth by this approach. In fact, there is a trade-off between the C_I/C_B capacitance ratio and the bandwidth of the buffer. Furthermore, in an n-well technology, NMOS source followers are the fastest, but are affected by the body effect, providing a considerable attenuation and level shifting (Säckinger and Fischer 2000).

4.1.2.2 Inverse Scaling

Another approach to reduce the equivalent capacitance of an inter-stage node is the inverse scaling (Jindal 1987). It consists of designing the width of the input transistors of the second stage k times smaller than the first one, as shown in Fig. 4.4.

Thus, the input capacitance of the second stage can be approximated, neglecting the wiring capacitance, as:

$$C_I' = C_I/k \tag{4.9}$$

Similar to the previous technique, the bandwidth is enhanced according to the expression:

$$BW' = \frac{C_I + C_O}{C_I/k + C_O} BW \tag{4.10}$$

In this case, the drawbacks associated with the included buffer, such as no signal attenuation, no new poles limiting the bandwidth, and no level shifting, are avoided. Hence, this technique is very popular for CMOS designs. However, if the post-amplifier is formed by several cascaded stages, the width ratio between the

Fig. 4.5 Miller effect

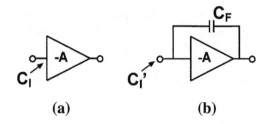

(a)   (b)

first and the last stage might grow considerably. For instance, for a four-stage post-amplifier and a $k = 2$ inverse scaling, the output stage becomes 16 times smaller than the input stage. Thus, a scaling ratio of 2 or even lower is a typical value to accomplish the following two conditions: Driving the output capacitance and the maximum allowable input capacitance. If this technique is applied only between two stages, as in our case, the scaling ratio may be higher.

4.1.2.3 Negative Capacitances

Negative capacitances technique is based on Miller effect (Miller 1920), which defines the influence of the feedback capacitance C_F on the input capacitance C_I, as illustrated in Fig. 4.5.

C_I is the input capacitance of the amplifier with no feedback capacitance. It must be noted that the feedback capacitance is formed by parasitic capacitances—C_{gd} capacitance for the conventional common source amplifier—and any feedback capacitor placed on purpose. Thus, the input capacitance including the Miller effect can be written as:

$$C_I' = C_I + (1 + |A|)C_F \qquad (4.11)$$

where A is the gain of the inverting amplifier. Thus, the contribution of the C_F capacitance to the input capacitance of the amplifier is multiplied by the gain due to the negative feedback loop, and hence, although the capacitance is small, its contribution might be significant and increase the input capacitance. For a positive feedback loop, the sign within parenthesis turns negative, reducing the input capacitance. Such a property might be exploited to enhance the bandwidth.

Single gain stages are usually inverting amplifiers, requiring two cascaded ones to create a non-inverting amplifier, and hence, a possible positive feedback. However, for a differential gain stage, this technique can be implemented as shown in Fig. 4.6.

Thus, the input capacitance of the differential stage with feedback capacitances can be written as:

$$C_I' = C_I + (1 - |A|)C_F \qquad (4.12)$$

It must be noted that for a gain stage $(A > 1)$, the Miller capacitance is negative, reducing the equivalent capacitance of the input node, and hence, enhancing the bandwidth (Tsai and Chen 2007).

Fig. 4.6 Differential stage **a** without and **b** with negative Miller capacitances

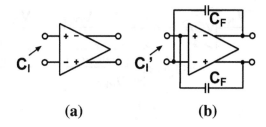

(a)   (b)

A limitation of this technique is that the feedback capacitors present an additional load to the output of the stage, which might reduce the bandwidth. In addition, the stability of the structure must be ensured, because if the feedback capacitors are made too large, the overall capacitance at the input stage may become negative (Razavi 2008).

4.1.2.4 Zero-Pole Cancellation

The last implemented technique is denominated the zero-pole cancellation. It is based on the complementary frequency effect of poles and zeros. The frequency response of a structure with dominant pole approximation can be expressed by the transfer function as:

$$H(s) = \frac{H_0}{s/BW + 1} \tag{4.13}$$

where s is the complex angular frequency, BW is the bandwidth (the value of -3 dB frequency), and H_0 is the value of the transfer function at low frequencies. If somehow we could introduce a zero at the same frequency of the pole, the bandwidth will be enhanced, represented by the transfer function as:

$$H(s) = \frac{H_0}{s/BW + 1} \left(\frac{s/BW + 1}{s/BW' + 1} \right) = \frac{H_0}{s/BW' + 1} \tag{4.14}$$

Therefore, the bandwidth will now be limited by higher frequency poles. To attain such a theoretical improvement for the bandwidth, there are no practical rules, and finding the way depends only on the designer.

4.2 Automatic Gain Control

To enhance the input dynamic range, it is necessary to fix the appropriate gain depending on the signal level (Israelsohn 2002). This is done by using an AGC circuit, which will choose high gain for low input signal, achieving high signal-to-noise ratio, and low gain for high input signal to avoid saturation.

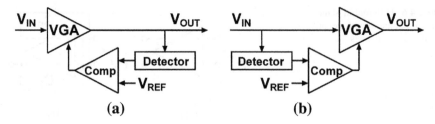

Fig. 4.7 Simplified block diagrams of **a** feedback and **b** feedforward AGCs

From a practical point of view, the most general description of an AGC system is presented in Fig. 4.7. The input signal V_{IN} is amplified by a variable gain amplifier (VGA), whose gain is controlled by a signal V_C. To adjust the gain of the VGA to its optimal output level V_{OUT}, the AGC generally first detects the strength level of the signal using the peak detector; it then compares this level with a reference voltage V_{REF} and finally, filters and generates the required control voltage.

This function can be performed by detecting the signal at the output of the VGA, and hence, the architecture is called "feedback" AGC (Fig. 4.7a), or at the input, in which case it is identified as "feedforward" AGC (Fig. 4.7b) (Alegre et al. 2011). Both the structures present different inherent characteristics that lead to choosing one or the other depending on the target application.

The advantages of using feedback AGC are as follows: First, the dynamic range required at the detector input is reduced in the same way as the AGC gain range. Second, the circuit linearity is high due to the feedback loops' inherent characteristic. On the other hand, this architecture also has some disadvantages. The high level of feedback required to reach high compression ratios makes the feedback processors more likely to exhibit instabilities if high compression ratios are managed. Instability is also likely in feedback expanders where high expansion ratios are desired. Finally, the feedback loop will always have a maximum boundary bandwidth to maintain stability. This maximum bandwidth entails a minimum settling time (Green 1983). Moreover, to keep the settling time constant, the feedback configuration requires the use of specific control voltage generation functions (Khoury 1998).

Feedforward AGC offers a time constant that mainly depends on the peak detector response; hence, this loop is ideally not affected by the minimum settling time restriction. In addition, high compression and high expansion ratios are possible with this configuration (Israelsohn 2002). In contrast, the disadvantages of a feedforward AGC are that the level detector is exposed to the entire dynamic range of the input signal and that the loop requires higher linearity, because the linearity improvement inherent to the feedback loop is now absent.

The proposed AGC amplifier is implemented by a (PGA) controlled by a digital word, instead of a VGA tunable by a continuous electrical variable. Thus, the main drawbacks of the feedback topology, such as instabilities for high compression ratios and settling time restriction, are avoided. The robust digital word controlling the gain of the amplifier helps to avoid instability in spite of high compression ratios. Finally, the settling time is not a drawback itself, because the envelope of a typical data signal changes much more slowly than the bit rate. As a result, feedback topology is chosen in this case due to its inherent advantages.

4.2.1 Linear-in-dB Gain Distribution

In addition to the maximum and minimum gain values selected for the amplifier, there is another important choice to make—the gain distribution. It can be demonstrated that the settling time of an AGC amplifier can be constant (Khoury 1998), independent of the amplifier gain, if the appropriate gain distribution is selected. Let us start the demonstration from a simplified AGC diagram, shown in Fig. 4.8.

The following two expressions can be derived from Fig. 4.8:

$$V_{OUT}(t) = G(V_C(t))V_{IN}(t) \tag{4.15}$$

$$V_C(t) = \int_0^t (V_{REF} - k\ln(V_{PEAK}(x)))dx \tag{4.16}$$

The first one is the evident relationship between the input and output signal, while the second shows the control voltage V_C generated by a typical AGC loop, consisting of the integration of the detected peak voltage of the output signal. An amplifier k and a reference voltage V_{REF} are usually included to generate the appropriate control voltage. The logarithmical block can be omitted; however, this theoretical demonstration becomes an approximation (Khoury 1998). Without losing generality, signals can be only represented by its amplitude. By choosing this particular logarithmical scale, we can obtain:

$$A_{IN,OUT,PEAK}(t) = \ln(Amp(V_{IN,OUT,PEAK}(t))) \tag{4.17}$$

$$A_{PEAK}(t) = A_{OUT}(t) \tag{4.18}$$

It can be noted that now there is no distinction between the amplitude of the output signal and the detected peak signal. With this nomenclature (4.15) can be written as:

$$A_{OUT}(t) = A_{IN}(t) + \ln G(V_C(t)) \tag{4.19}$$

and taking its derivate with respect to the time and introducing the expression of the control voltage from (4.16), the following equation is obtained:

$$\frac{dA_{OUT}(t)}{dt} = \frac{dA_{IN}(t)}{dt} + \frac{1}{G(V_C(t))}\frac{dG(V_C(t))}{dV_C}(V_{REF} - kA_{PEAK}(t)) \tag{4.20}$$

Equation (4.20) describes a first-order linear system having a high pass response with a time constant given by:

Fig. 4.8 Simplified AGC loop block diagram

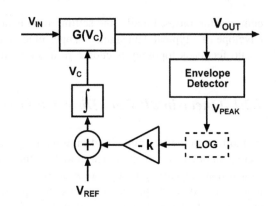

$$\tau = \frac{k}{G(V_C)}\frac{dG(V_C)}{dV_C} \tag{4.21}$$

Therefore, the time constant is directly related to the gain distribution. To attain a time constant independent of the gain variation, an exponential distribution—in other words, linear-in-dB—must be achieved.

$$\tau = k \cdot k_2 = const. \Leftrightarrow G(V_C) = k_1 e^{k_2 V_C} \tag{4.22}$$

To illustrate the theoretical demonstration, the time responses for two different gain distributions—linear and linear-in-dB—are represented in Fig. 4.9.

4.2.2 Discrete Gain Distribution

As previously mentioned, an AGC amplifier design was selected for post-amplifier, instead of a limiting amplifier, obtaining constant output amplitude over the whole dynamic range of the receiver. In fact, the constant output amplitude is not a strict requirement; the only strict requirement is that the output signal has sufficient amplitude level (Muller and Leblebici 2007). Accordingly, a PGA was selected to implement the post-amplifier, attaining a limited output amplitude range with a more robust digital gain control. Thus, a discrete linear-in-dB gain distribution was implemented.

We must remark that the previous demonstration is valid for a VGA, but not for a PGA, due to the use of the derivative in (4.20), which requires a continuous function $G(V_C)$. However, attaining a linear-in-dB gain distribution for a PGA is as or even more important than that for a VGA. Such a gain distribution provides a constant gain ratio between two consecutive gain states G_n and G_{n+1}:

$$\frac{G_{n+1}}{G_n} = \Delta G = const. \tag{4.23}$$

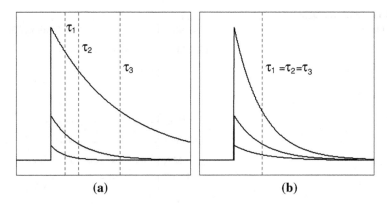

Fig. 4.9 Time responses depending on gain distribution: **a** linear and **b** linear-in-dB

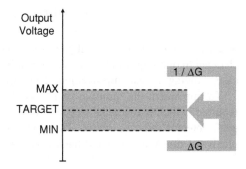

Fig. 4.10 Gain variation
diagram

Thus, the gain step ΔG does not depend on the gain state, n, attaining a particular gain variation diagram, defined by a maximum and a minimum possible value for the output amplitude, which is illustrated in Fig. 4.10.

To implement this basic gain variation diagram, only a peak detector and two comparators (for MAX and MIN values) are required in addition to the digital circuitry. However, synchronous sequential systems with several states offer different settling time, depending on the number of states to be changed. In addition, the worst-case settling time—from the first to the last state or backwards—might become too long. To overcome these drawbacks, several states can be changed in one step. Figure 4.11 illustrates the case with the possibility of double step. Four levels (MAX, MIN, $\Delta G \cdot$MAX, and MIN$/\Delta G$) must be defined to decide the most appropriate gain step to attain the proper amplitude range.

Thus, the maximum number of steps $n_{\max}$ to reach the desirable gain state and the probability P to do this in one single step are given by:

$$n_{\max} = \text{ceil}\left(\frac{n_T - 1}{s}\right) \tag{4.24}$$

Fig. 4.11 Improved gain
variation diagram

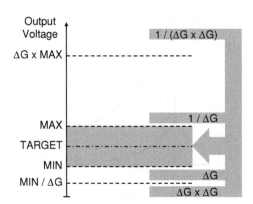

$$P(\%) = 100 \frac{n_T + \sum\limits_{j=1}^{s} 2(n_T - j)}{n_T^2} \tag{4.25}$$

where n_T is the number of total gain states and s is the number of states that can be changed in one step. Both the parameters are illustrated in Fig. 4.12 for a particular case showing a clear improvement depending on the step size.

To implement the multiple-step topology, a higher number of comparators is required, which is twice the step size s. Moreover, the input dynamic range of the peak detector must be wider. Therefore, there is a trade-off between the complexity of the AGC loop and the worst-case settling time.

4.3 Offset Compensation

The third basic component of an AGC is the DC offset compensation circuit. For high-gain amplifiers, DC offset compensation is mandatory to keep the amplifier DC output voltage approximately constant, in spite of process-voltage-temperature (PVT) variations (Crain and Perrot 2006). If it is not implemented, a small input offset can be hugely amplified degrading the output signal, as illustrated in Fig. 4.13. Furthermore, input offset increases the amplitude detected (V_{PD}) by the peak detector as shown in Fig. 4.14, and consequently, the selected gain by the AGC circuit might not be the most appropriate for the input signal level.

Therefore, the error caused in the peak voltage (ΔV_{PD}) can be calculated as:

$$V_{PD} = \max(V_1, V_2) + V_S/2 \Rightarrow \Delta V_{PD} = \max(V_1, V_2) - V \tag{4.26}$$

As shown in Fig. 4.15, a typical offset compensation circuit is formed by an integrator, which can be as simple as an RC filter, and a non-inverting amplifier A_0 to close a negative loop for an inverting core (García del Pozo 2010). For a differential implementation, this circuit must be duplicated. Therefore, only lower frequency signals, in particular, DC signal, are affected.

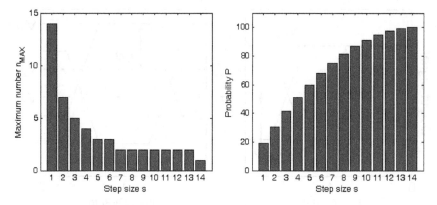

Fig. 4.12 Characteristics of sequential distribution depending on step size for $n_T = 15$

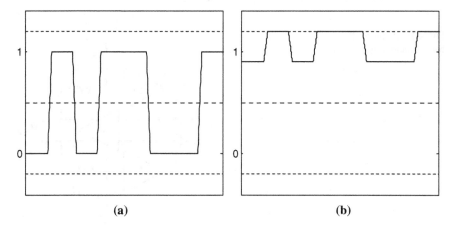

Fig. 4.13 Signal amplified without compensation **a** ideally and **b** affected by input offset

The effect of the compensation loop is visually illustrated in terms of the frequency response shown in Fig. 4.16. Accordingly, two new parameters can be determined. First, a low cut-off frequency f_{LF} appears, affecting the sensitivity, as explained in Sect. 2.4.3. The sensitivity penalty, presented in Sect. 2.4.3, can be minimized by reducing the low cut-off frequency (Maxim Integrated Products 2008). On the other hand, the desired gain reduction ΔG is achieved for lower frequency signals, minimizing the output offset.

$$\Delta G = A_0 A + 1 \tag{4.27}$$

$$f_{LF} = \frac{A A_0 + 1}{2\pi CR} \tag{4.28}$$

As seen, both the parameters depend on the post-amplifier gain A. Thus, when the gain of the PGA is reduced, the offset correction is also reduced. However,

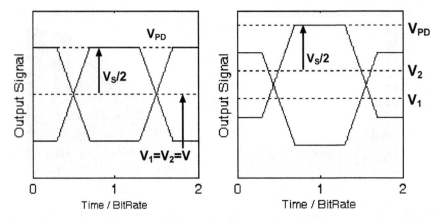

Fig. 4.14 Detected peak value **a** ideally and **b** affected by input offset

Fig. 4.15 Typical offset
compensation circuit

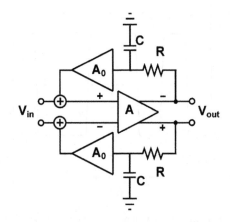

Fig. 4.16 Frequency
response with offset
compensation circuit

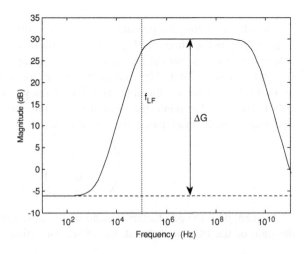

lower gain states are only active with higher input signals, and thus, they are less affected by the same input offset voltage. Therefore, the most critical state—highest gain—is also the most influenced by the offset compensation technique.

Following the introduction of the theoretical analysis of the post-amplifier, the implementation of a CMOS circuit, including all the mentioned aspects, is detailed in the subsequent sections.

4.4 Proposed AGC Design

In this section, an AGC amplifier targeting multi-gigabit transmission is presented. It consists of a four-stage amplifier core, offering a linear-in-dB discrete gain distribution with seven gain states, an AGC loop based on a double-step topology to minimize the settling time, and the mandatory offset compensation circuit. First, the design of the amplifier core—selected gain stage, broadband techniques, and programmable gain technique—is detailed. Then, the AGC loop implementation, including a peak detector, four comparators, and a shift register, is presented. Finally, the offset compensation circuit is analyzed.

4.4.1 Amplifier Core Architecture

The design of the amplifier core is the main challenge within the post-amplifier design, due to the required high bandwidth (Hermans and Steyaert 2005). A fully differential structure was adopted due to the higher immunity to environmental noise. As the implemented TIA, the amplifier core is designed limiting the maximum number of transistors between supply and ground, and thus, it will be suitable for downscaled CMOS technologies or a reduction in the power dissipation by reducing supply voltage may be explored.

4.4.1.1 Differential Gain Stage

Gain stages are based on the common-mode feedforward pseudo-differential pair introduced by Snelgrove and Shoval (1992). As shown in Fig. 4.17, it consists of an NMOS pseudo-differential pair, N_1 and N_2, biased through the PMOS current mirror $P_1 - P_2$. The input common-mode voltage will be suppressed due to the subtraction at the output of the current provided by the transistor N_1 and the mirrored current provided by N_2, while a differential input voltage will cause an output current I_{OUT}. This gain cell was selected, instead of the more conventional differential gain cell, to explore new possibilities that may lead to low-voltage low-power operation.

For this simple topology, considering perfectly matched transistors $N_1 - N_2$, the transconductance can be expressed as:

Fig. 4.17 Snelgrove cell

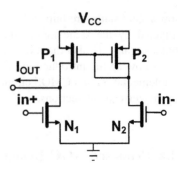

$$G_m = \frac{I_{OUT}}{V_{IN}} = \mu_N C_{OX} \frac{W_N}{L_N} (V_{CM} - V_{TH}) \tag{4.29}$$

where C_{OX} is the oxide capacitance of MOS devices; V_{TH} and μ_N are $N_1 - N_2$ threshold voltage and mobility, respectively, which are technological parameters; V_{CM} is the input common-mode bias voltage; and W_N and L_N are $N_1 - N_2$ width and length, respectively, which are design parameters. To optimize the frequency performance of the cell, the length of the MOS was chosen to be minimal. As the transconductance cell provides a single-ended output, such a cell must be duplicated to obtain a fully differential structure (Calvo et al. 2008), shown in Fig. 4.18.

The output currents of this transconductance, G_m, cell are converted to voltage through polysilicon resistive loads R_L. Thereby, the differential gain of this stage can be written as:

$$G = G_m \cdot R_L \tag{4.30}$$

where G_m is the differential transconductance in (4.29). A value of $R_L = 1$ kΩ was chosen to achieve a good gain, bandwidth, and power trade-off. The common-mode voltage V_{CM} is set to 900 mV, a half of the supply voltage. Therefore, to generate the same DC common-mode output voltage to allow direct coupling, a constant current, I_B, implemented with PMOS biased in saturation, was added to each output node.

4.4.1.2 Fourth Stage

The fourth stage of the amplifier may be considered as the first step to enhance the overall bandwidth (Aznar et al. 2008), implementing the inter-stage buffering technique. In this way, it must provide low input capacitance and low output impedance, making it possible to adapt the high output impedance (~ 1 kΩ) of the latest gain stage A_3 to the load capacitance of the subsequent stage, modeled by $C_L = 100$ fF. Without this buffer, we would have a pole $\omega_p \approx 1.5$ GHz in the frequency response, making it impossible to achieve the necessary bandwidth for multi-gigabit applications.

In CMOS technology, there are two conventional implementations of buffers, namely, common-source and common-drain (Säckinger and Fischer 2000). The

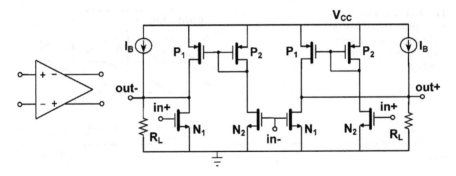

Fig. 4.18 Differential gain stage based on Snelgrove cell

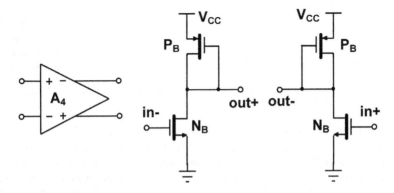

Fig. 4.19 Fourth stage schematic

latest one shows a considerable level shift, and hence, it is hardly suitable for low-voltage technologies. A common-source structure biased with a PMOS in diode was selected to implement A_4, as shown in Fig. 4.19, with low output impedance given approximately by (4.31).

$$R_O \approx \frac{1}{g_{m,P_B}} \qquad (4.31)$$

The transistor dimensions are 12/0.18 (μm/μm) for NMOS and 48/0.18 for PMOS. With these selected transistor sizes, the input capacitance and output impedance remain low, as they are proportional to $W_N \cdot L_N$ and L_P/W_P, respectively. On the other hand, the buffer has an attenuation of -3.7 dB and it provides an output DC voltage of 1 V.

4.4.1.3 Multi-Stage Amplifier Architecture

The proposed amplifier architecture (Aznar et al. 2008), shown in Fig. 4.20, is a fully differential amplifier consisting of three cascaded gain stages, denoted by $A_1 - A_3$ in the figure, followed by another stage (A_4).

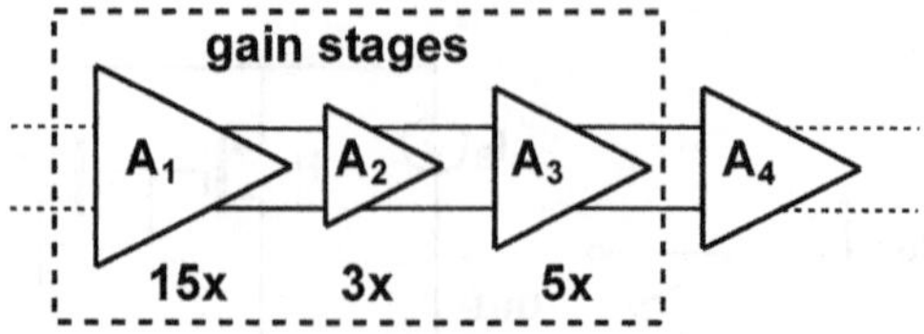

Fig. 4.20 Proposed multi-stage amplifier indicating scaling ratio among gain stages with lowest integer numbers

Table 4.1 Design parameters for amplifier core

Stage	Instance	Width/value
A_1	N_i	30 μm
	P_i	24 μm
A_2	N_i	6 μm
	P_i	4.8 μm
A_3	N_i	10 μm
	P_i	8 μm
A_4	N_B	12 μm
	P_B	48 μm
$A_1 - A_3$	I_B	900 μA
	R_L	1 kΩ

To optimize the bandwidth, inverse scaling is implemented between the first two stages. As a result, the three gain stages are not identical; they differ in the width of the input NMOS pair, W_N, and therefore, in the gain that they exhibit. The first gain stage A_1 has the lowest impedance input node. Thus, we chose the largest $W_N = 30$ μm for A_1, which resulted in the highest gain value of 16.9 dB. According to an inverse scaling rate of 5, $W_N = 6$ μm was selected for A_2, which resulted in a gain value of 8.4 dB. As it was impossible to continue with the same inverse scaling rate, we designed A_3 with a larger width than the second stage, $W_N = 10$ μm, obtaining the desired gain value of 11.4 dB. As a result, we had an overall maximum gain, including the attenuation of the last stage, A_4, of 33 dB. Table 4.1 summarizes the main design values of all stages of the amplifier core.

4.4.1.4 Bandwidth Enhancement

To further enhance the bandwidth, a resistor R of 890 Ω was included in the fourth stage, as depicted in Fig. 4.21. First, if we analyzed the simplified AC model of this stage, including the load capacitance associated with the next stage, as shown in Fig. 4.22; when the resistor $R = 0$, the transfer function is:

$$H(s) = \frac{V(s)}{I(s)} = \frac{1}{g_m + s(C_L + C_{gs})} \qquad (4.32)$$

where g_m and C_{gs} are the transconductance and the gate-source capacitance (60 fF) of the biasing PMOS P_B, respectively. When the resistor R is included, the transfer

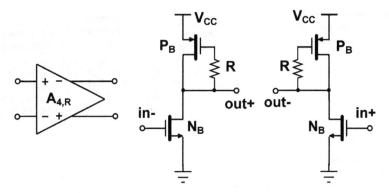

Fig. 4.21 Fourth stage with a resistor to enhance the bandwidth

Fig. 4.22 Small signal
equivalent model of fourth
stage

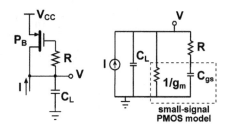

function in (4.32) is modified, and it exhibits a zero and two poles according to the expression:

$$H(s) = \frac{1 + sC_{gs}R}{s^2 RC_{gs}C_L + s(C_L + C_{gs}(1 + g_mR)) + g_m} \qquad (4.33)$$

If $g_m R \gg 1$, zero-pole cancellation occurs and bandwidth increases:

$$H(s) \approx \frac{1 + sC_{gs}R}{(1 + sC_{gs}R)(g_m + sC_L)} = \frac{1}{g_m + sC_L} \qquad (4.34)$$

In addition to this approach, the negative capacitances technique was used at the third stage: The input capacitance was compensated with a cross-coupled capacitance C_F pair in a positive feedback loop configuration. These C_F capacitors of only 30 fF were implemented with dummy PMOS transistors to obtain a good matching between input capacitance and C_F (McCreary 1981).

Figure 4.23 illustrates the proposed post-amplifier, attaining 33 dB gain and exceeding 3 GHz bandwidth, owing to the applied broadband techniques (Aznar et al. 2008). The details of the simulated improvement of bandwidth for each technique are illustrated in Fig. 4.24 and Table 4.2. Although the results might suggest that some techniques are preferable, the required bandwidth is only achieved by the combination of all the applied broadband techniques.

Fig. 4.23 Post-amplifier
architecture indicating all
broadband techniques applied

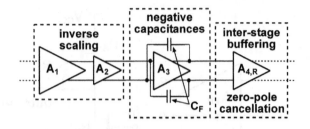

Fig. 4.24 Simulated
bandwidth enhancement of
the post-amplifier

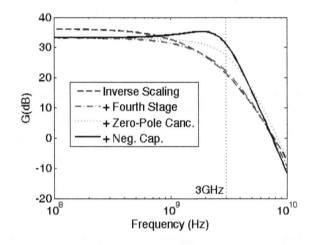

Table 4.2 Simulated
bandwidth depending on the
broadband techniques applied

Broadband technique	Bandwidth (GHz)	Improvement	
Inverse scaling	0.9	–	–
Fourth stage	1.3	0.4	44.4%
Zero-pole cancellation	2.2	0.9	69.2%
Negative capacitances	3.1	0.9	40.9%

4.4.2 Programmable Gain

The gain of a differential common-source amplifier is, as in our proposed gain cell
(4.30), the product of a transconductance and a load resistor. As a result, the gain
can be modified by varying one of these parameters. It must be remarked that the
selected gain variation technique must not modify the output common-mode, as
the operating point of the subsequent gain stage would vary.

The most popular gain variation technique is the degenerated source. However,
it is not suitable for our proposed cell because there is no source bias current. The
selected technique to change the gain is the variable load resistor to facilitate an
optimized layout of the Snelgrove cell, which is illustrated in Fig. 4.25 for a
differential common-source amplifier.

Fig. 4.25 Differential gain stage including variable load resistor

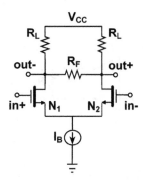

$$G = g_m R_{EQ} = g_m \left(R_L \left\| \frac{R_F}{2} \right. \right) \qquad (4.35)$$

As shown in (4.35), the gain is modified due to the equivalent output resistance R_{EQ}, while the transconductance of the NMOS transistors g_m remains invariant. A floating load resistor R_F was chosen to modify the gain without affecting the operation point of the transistors and the common-mode output voltage. The main drawback of this technique is an enhancement of the bandwidth as the gain is reduced; however, the bandwidth of the whole receiver remains approximately constant as it is dominated by the TIA. In addition, this effect will be partially compensated by the broadband techniques implemented. This technique offers a constant input dynamic range and a good noise performance, while reduction of the transconductance usually affects the input dynamic range and worsens the noise performance. Figure 4.26 illustrates the proposed gain cell with the selected gain variation technique.

$$G = G_m R_{EQ} = G_m \left(R_L \left\| \frac{R_F}{2} \right. \right) \qquad (4.36)$$

With regard to the previous case, the gain is reduced by implementing a floating load resistor R_F, while the transconductance of the Snelgrove cell G_m remains invariant. The typical implementation of a variable resistor in CMOS technology is a transistor biased in triode region. Thus, such a transistor can be modeled by a resistor controlled by the gate voltage.

To attain a discrete gain distribution, i.e., a PGA, an NMOS array was selected to implement R_F (Sanz et al. 2005), as shown in Fig. 4.27. Thus, a digital word $D \equiv \{b_1...b_n\}$ controls the value of the equivalent resistance and, hence, the gain of the stage. The highest gain state is achieved with all bits OFF, while from each bit ON, the load resistor at the output of the stage is reduced, reducing the gain of the stage.

In detail, as illustrated in Fig. 4.28, R_F was implemented by an array of six NMOS to control gain A_1 by the digital word $D \equiv \{b_1...b_6\}$ and three NMOS for gains A_2 and A_3, which are controlled by $\{b_1b_2b_3\}$ and $\{b_4b_5b_6\}$, respectively. Thus, the gain control is shared among the three gain stages, extending the gain range.

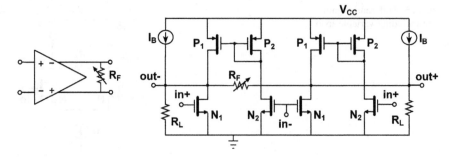

Fig. 4.26 Proposed gain stage including variable load resistor

Fig. 4.27 Implementation of R_F with a NMOS array

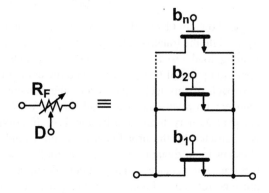

Fig. 4.28 Proposed NMOS arrays for the three gain stages

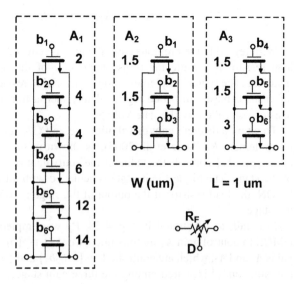

Fig. 4.29 Simulated gain distribution

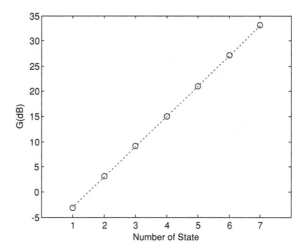

Fig. 4.30 Gain variation diagram

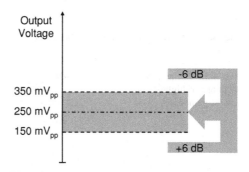

In particular, through a thermometer code control, the amplifier showed a linear-in-dB gain distribution. Consequently, a constant settling time of the AGC loop was achieved (Khoury 1998). As shown in Fig. 4.29, the post-amplifier demonstrated a gain distribution from 33 to −3 dB in a step of 6 dB, which led to defining two levels limiting the output amplitude range, as shown in Fig. 4.30.

Therefore, the desired 250 mV peak-to-peak output voltage was transformed to output amplitude of 150–350 mV peak-to-peak, showing sufficient amplitude over the whole range. When the output amplitude is out of this range, the AGC loop must change the gain to remain within the range, as shown by the arrows in Fig. 4.30.

The impact on the performances of the post-amplifier due to the gain variation must be studied now, especially because the bandwidth, as the main drawback of the selected technique to change the gain, shows a considerable variation. The post-amplifier requires a bandwidth exceeding 3 GHz over the whole gain range and a constant bandwidth when varying the gain is desired. Furthermore, peaking in the frequency response must be avoided. The simulated frequency response is shown in Fig. 4.31.

Both the requirements could be achieved owing to the gain control being shared out among the three gain stages. In addition, an almost constant bandwidth was

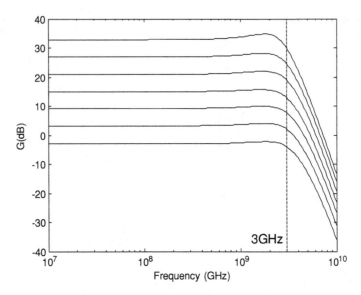

Fig. 4.31 Frequency response over the whole gain range

Fig. 4.32 Frequency response of highest and lowest gain state

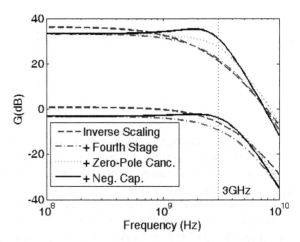

obtained due to the compensation of two opposite effects: The bandwidth is enhanced as the overall gain is reduced due to the selected gain variation technique, and the negative capacitances technique increases the bandwidth depending on the gain of stage A_3, as indicated in (4.12). A study of the bandwidth enhancement is shown in Fig. 4.32 and Table 4.3.

Simulation results show an almost constant bandwidth because of no effect of negative capacitances technique for lowest gain state, as the input capacitance compensation (4.12) caused by this technique does not occur when the gain of A_3 stage is reduced.

Table 4.3 Simulated bandwidth for highest and lowest gain state

Broadband technique	Highest gain state			Lowest gain state		
	BW (GHz)	Improvement		BW (GHz)	Improvement	
Inverse scaling	0.9	–	–	1.8	–	–
Fourth stage	1.3	0.4	44.4%	2.0	0.2	11.1%
Zero-pole cancellation	2.2	0.9	69.2%	3.6	1.6	80%
Negative capacitances	3.1	0.9	40.9%	3.6	0	0%

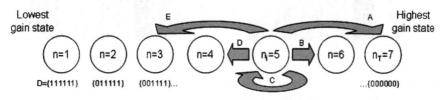

Fig. 4.33 Proposed state diagram

4.4.3 AGC Loop

For feedback AGC loops, a sequential state diagram is the simplest implementation. Therefore, only two levels are required to determine when the gain must be changed, as shown in Fig. 4.30. It must be noted that this is only possible if a linear-in-dB gain variation is implemented, as presented in Sect. 1.2.1.

In this section, the design and implementation of an AGC loop, including the possibility of a double step, is detailed (Aznar et al. 2009). The settling time of the amplifier is improved due to the reduction in the maximum number of steps n_{max} to reach the desirable gain state (4.24) and the increase in the probability P to do this in one single step (4.25).

4.4.3.1 Proposed State Diagram

In the proposed design (Fig. 4.33), the 6-bit thermometer gain code determines seven gain states, and hence, a step size $s = 2$ was chosen to reduce n_{max} (50 %) and improve P (20.4 %), when compared with a pure sequential version ($s = 1$), thus optimizing the settling time without a high increase in the complexity of the circuit. Both the parameters are represented in Fig. 4.34 for this particular case.

The gain state numbered 5 was selected as the initial state (n_i in Fig. 4.33) because the amplifier has more likelihood to require a high gain state, while any state can be reached in only two steps. Thus, the gain variation diagram, shown in Fig. 4.35, was improved with two new possible steps that could help to reach the necessary gain state in case a huge signal variation happens suddenly. Now, four levels are required to determine when the gain must be changed, substantially improving the simplest sequential implementation.

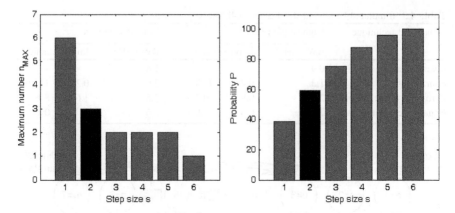

Fig. 4.34 Characteristics of sequential distribution depending on step size for this case ($n_T = 7$), with the selected step size highlighted in black

Fig. 4.35 Proposed gain variation diagram

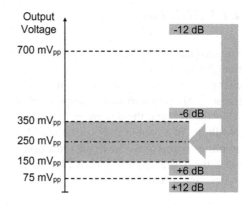

4.4.3.2 AGC Loop Implementation

In accordance with the above-mentioned factors, the AGC feedback loop was formed by a peak detector, four comparators, and a shift register, as shown in Fig. 4.36. Briefly, the peak detector provides a voltage proportional to the output signal amplitude, which is processed using four comparators to establish five gain-setting possibilities, namely, (1) no change for gain, (2) a gain increase of +6 dB, (3) a gain increase of +12 dB, (4) a gain reduction of −6 dB, or (5) a gain reduction of −12 dB. Thus, according to the output of the comparators, the 6-bit thermometer gain code control—kept in a shift register—is held or modified. The complete AGC loop design is described in the following paragraphs.

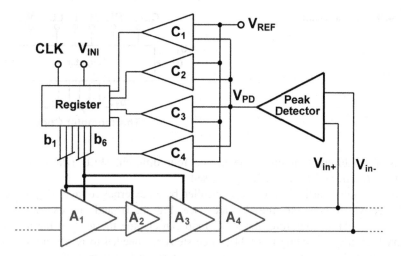

Fig. 4.36 AGC loop block diagram

Fig. 4.37 Generic double-
direction shift register

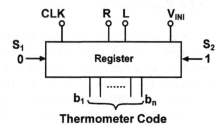

4.4.3.3 Shift Register

An n-bit thermometer code control can be implemented directly by an n-bit double-direction shift register (Gajski 1997). All synchronous sequential circuits require two inputs, namely, a clock signal (*CLK*) and an initializing signal (*V$_{INI}$*), because the output depends on their previous state. Digital elements usually have a reset (all outputs LOW) or set (all outputs HIGH) as the initializing signal. In this case, the initial output digital word might be of any consistence with the thermometer code. The LOW to HIGH transitions of clock signal establishes when the output digital word can be changed.

In addition to the required inputs, the operation of a double-direction shift register is controlled by four inputs. Two inputs determine the direction in which the digital word is shifted (*R* for right and *L* for left), while serial inputs (*S$_1$* and *S$_2$*) provide the necessary input bit depending on the shift direction. Serial inputs must also be consistent with the thermometer code. Figure 4.37 illustrates the shift register.

The designed shift register, shown in Fig. 4.38, is used to store and modify the 6-bit digital word *D* that fixes the gain implementing the proposed state diagram.

Fig. 4.38 Designed shift
register

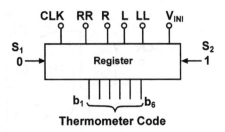

Thus, four different inputs to determine four different shifts (*RR* twice right, *R* right, *L* left, and *LL* twice left) are used.

As explained earlier, it must be controlled by an external voltage, V_{INI}, which is only active to set the initial state (see Fig. 4.33) and an external clock ($f_{CLK} = 4$ MHz). Therefore, the AGC loop can change the gain of the amplifier every 0.25 μs. According to (4.34), three steps are needed to reach the desirable gain, which establishes a worst-case settling time of less than 1 μs.

4.4.3.4 Peak Detector

The peak detector structure, shown in Fig. 4.39, is a differential positive scheme that implements a full-wave rectifier circuit through a transconductor, followed by a unidirectional current mirror (Park et al. 2006). When the input V_{in} is larger than the output $V_{PD,}$ the excess current flowing in the current mirror charges a hold capacitor C_D. A differential structure has been considered to help maintaining the peak value, avoiding discharges of C_D between maximums of the detected signal amplitude, as shown in Fig. 4.40.

When the input is smaller than the output, the current source I_D slowly discharges C_D. Subsequently, the peak detector becomes an envelope detector (Alegre et al. 2008). To follow the amplitude of the signal in any case, a faster discharge of the capacitor must be implemented. If not, a large reduction in the output signal will not be followed by the detector. The current source is formed by two NMOS transistors biased, when ON, in saturation region, as shown in Fig. 4.41.

The MOS transistor N_S is always ON, implementing the slow discharge mode. When the switch, implemented by a transmission gate, is close, transistor N_F, designed wider than N_S, also works in the saturation region, increasing the discharging current. Thus, the discharging effect, quantified by the slope, is modified:

$$slope = \frac{I_D}{C_D} \qquad (4.37)$$

In particular, with an implemented MIM capacitor of 1 pF, the simulation results show a slope of 0.2 and 3.8 mV/ns for slow and fast mode, respectively. The main design values of the peak detector are summarized in Table 4.4.

The fast discharge mode is active only when two conditions are fulfilled. First, *CLK* must be HIGH. Thus, the fast discharge mode is not active when the digital

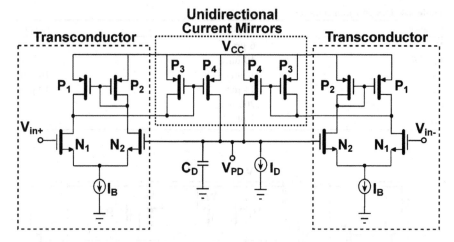

Fig. 4.39 Peak detector schematic

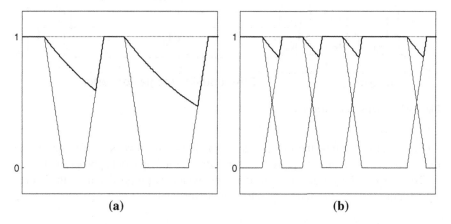

Fig. 4.40 Detected peak value for **a** single-ended and **b** differential signal

Fig. 4.41 Slow and fast
discharge structure

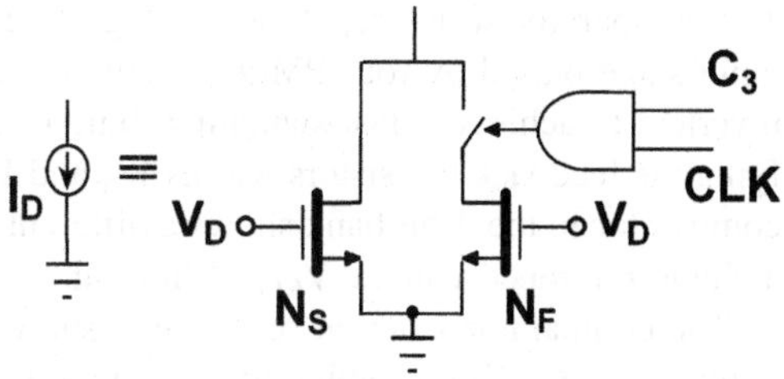

word is about to be changed, avoiding large ripple in the detected peak value
produced by a high discharge current. In addition, the output of comparator number

Table 4.4 Design parameters for peak detector

Instance	Width/value	Length
$N_1 - N_2$	6 µm	180 nm
$P_1 - P_4$	10 µm	180 nm
N_S	1 µm	10 µm
N_F	14 µm	10 µm
I_B	200 µA	
C_D	1 pF	

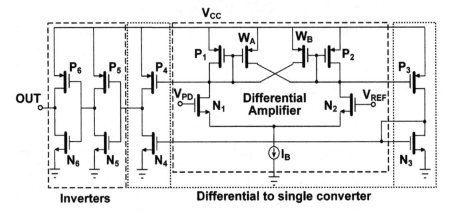

Fig. 4.42 Comparator schematic

3 must be HIGH. Thus, the fast discharge mode will be always active after a gain reduction, ensuring that the peak detector is discharged up to an appropriated level.

To sum up, the output of peak detector, which is proportional to the output signal amplitude, is carried to a four comparator bank.

4.4.3.5 Comparators

The comparator structure, shown in Fig. 4.42, consists of a differential amplifier input stage biased by four PMOS, a differential to single converter, and two final inverters to achieve a fast-switching digital output. With this structure, by modifying the feedback transistors widths W_A and W_B of the differential amplifier, each comparator in the 4-bit bank shows a different decision level, while only requiring a single reference voltage V_{REF} (Lin et al. 2004).

The comparator's output response is shown in Fig. 4.43. The four comparator outputs [C_1, C_2 C_3 C_4] define five regions (A, B, C, D, and E in Fig. 4.43), which correspond to five gain-setting possibilities: A gain step of +12, +6, 0, −6, or −12 dB, respectively, as illustrated in Fig. 4.33. These results are summarized in Table 4.5.

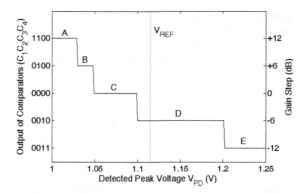

Fig. 4.43 Comparator output response

Thus, the comparator outputs $[C_1\ C_2\ C_3\ C_4]$ directly correspond to the inputs of the register $[RR\ R\ L\ LL]$ as long as R and L have no effect when RR or LL is active, respectively. The common parameters of all comparators are summarized in Table 4.6 and the particular widths W_A and W_B of each comparator are presented in Table 4.7. These values are limited by the width of $P_1 - P_2$ transistors, as the positive feedback must not dominate the behavior of the differential amplifier. The corresponding length L_A and L_B is 500 nm, equal to $P_1 - P_2$ transistors.

4.4.4 Offset Compensation Loop

As shown in Fig. 4.44, the proposed DC offset compensation circuit (Aznar et al. 2011) (one for each branch) consists of a simple RC integrator formed by MOS transistors to optimize the area, an inverting amplifier ($A_0 < 0$) that is implemented by three cascaded common-source transistors, and a PMOS current source transistor P_{OF}, also shown in Fig. 4.45, which adjusts the bias current I_B of the first gain stage A_1.

The negative closed loop is formed by the three inverting stages $A_2 - A_4$ of the core amplifier, the inverting amplifier A_0, and the inverting common source structure formed by P_{OF} and R_L. The low cut-off frequency introduced by this feedback loop has been fixed at 100 kHz, minimizing the sensitivity penalty caused by this effect as explained in Sect. 2.4.3. Furthermore, such a value of the low cut-off frequency is suitable for most of communication standards.

The silicon area of the offset compensation circuit is dominated by the RC integrator. Both passive elements are formed by MOS transistors to optimize the area. The resistance is implemented by a PMOS (P_{RES}) due to the lower mobility, while the capacitance is implemented by a NMOS (N_{CAP}) to connect drain and source to ground. The area is further optimized by a square layout of N_{CAP} (W/L = 30 μm/30 μm) and a "snake" layout of four transistors (W/L = 1 μm/ 15 μm) in series to implement P_{RES}.

Table 4.5 Gain changes possibilities

Output of comparator # (V)				Next gain step (dB)
C_1	C_2	C_3	C_4	
1.8	1.8	0	0	+12
0	1.8	0	0	+6
0	0	0	0	0
0	0	1.8	0	−6
0	0	1.8	1.8	−12

Table 4.6 Common design parameters for comparators

Instance	Width/value	Length
$N_1 - N_2$	3 µm	1.5 µm
$P_1 - P_2$	25 µm	500 nm
$N_3 - N_6$	3 µm	500 nm
$P_3 - P_6$	5 µm	500 nm
I_B	27.5 µA	

Table 4.7 Particular design parameters for each comparator

Comparator	W_A (µm)	W_B (µm)
C_1	22.5	–
C_2	22.5	4.5
C_3	22.5	16
C_4	–	22.5

Fig. 4.44 Implemented offset compensation circuit

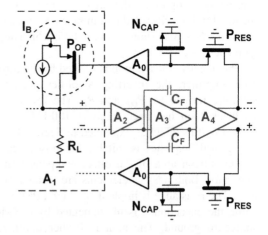

Figures 4.46 and 4.47 clearly illustrate the improvement in the output offset voltage of the complete amplifier when the offset compensation circuit is used for

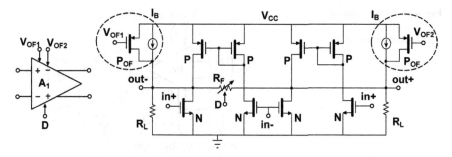

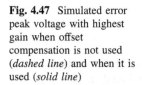

Fig. 4.45 Detail of first stage A_1 including offset compensation control

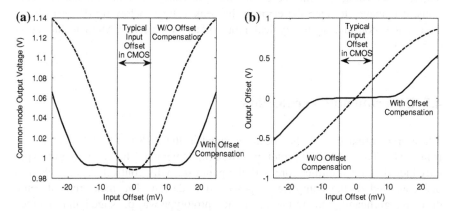

Fig. 4.46 **a** Output offset voltage and **b** common-mode output voltage of the complete amplifier with highest gain when offset compensation is not used (*dashed line*) and when it is used (*solid line*)

Fig. 4.47 Simulated error peak voltage with highest gain when offset compensation is not used (*dashed line*) and when it is used (*solid line*)

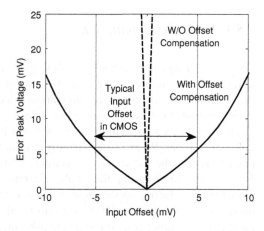

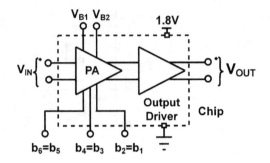

Fig. 4.48 Programmable post-amplifier implementation for PCB

the maximum amplifier gain setting (worst case). It minimizes the post-amplifier output offset voltage, while the common-mode voltage remains almost constant. Thus, it manages to keep the error in the peak voltage below 6 mV over all the gain range, considering the typical input offset voltage in CMOS technologies (±5 mV).

4.5 Experimental Verification

The aforementioned AGC has been designed in the 0.18 μm CMOS technology from UMC. Three different prototypes were implemented: The first two prototypes only included the amplifier core with the gain externally controlled, while the last one consisted of the complete proposed AGC design. The measurement setup differed significantly among them. The first prototype was packaged and mounted on a PCB, while the next prototypes were measured on-wafer to minimize parasitic effects.

4.5.1 PGA Implementation

To test the frequency response and the gain distribution of the proposed fully differential post-amplifier, the first prototype was implemented. It only includes the amplifier core, the proposed gain variation technique with external control, and a 50 Ω output driver, as shown in Fig. 4.48.

As the offset compensation circuit was not implemented, there were two bias voltages (V_{B1} and V_{B2}) to independently control the operation point of each signal path. To reduce the number of connections, a 3-bit digital word $\{b_1 = b_2, b_3 = b_4, b_5 = b_6\}$ was used, testing the whole gain range with a higher gain step (12 dB).

A further reduction in the number of connections was desired to facilitate the on-wafer characterization. Thus, a 2-bit digital word $\{a_2, a_1\}$ was used to determine the same number of states as the 3-bit thermometer code control.

Table 4.8 Digital control gain of programmable post-amplifier tested on-wafer

a_2	a_1	$b_6 = b_5$	$b_4 = b_3$	$b_2 = b_1$
0	0	0	0	0
0	1	0	0	1
1	0	0	1	1
1	1	1	1	1

Fig. 4.49 Programmable post-amplifier implementation tested on-wafer

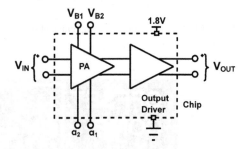

By examining Table 4.8, the following relationships were found:

$$b_6 = b_5 = a_1 \cdot a_2 \tag{4.38}$$

$$b_4 = b_3 = a_1 \tag{4.39}$$

$$b_2 = b_1 = a_1 + a_2 \tag{4.40}$$

Therefore, only two gates (OR and AND) must be implemented to reduce one connection. Figure 4.49 illustrates the PGA implementation. It must be remarked that although the PGA implementation might seem very similar in both the cases, the layout is completely different due to the strict PAD structure, which is mandatory to test on-wafer. In addition, on-chip resistors at the input are required to achieve 50 Ω matching.

4.5.2 AGC Implementation

The complete implementation of the proposed AGC system was also designed to be measured on-wafer. Therefore, the number of external connections must be minimized. Offset compensation and AGC avoid a double-bias voltage and the external digital word, respectively. However, AGC loop requires some external control voltages (V_{REF}, CLK, and V_{INI}), which leads to the conservation of the same number of external connections, as shown in Fig. 4.50. This implementation provides a full characterization of the proposed AGC post-amplifier, including the frequency response and gain distribution of all the gain states.

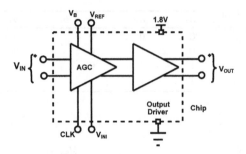

Fig. 4.50 AGC post-amplifier implementation

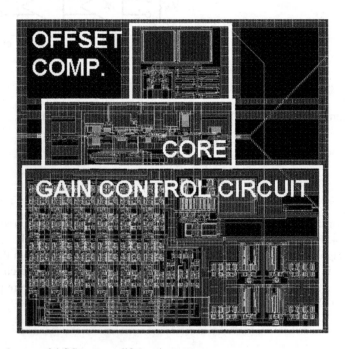

Fig. 4.51 Layout of AGC post-amplifier active area

As an example of the implemented prototypes, Fig. 4.51 shows the active area of the AGC post-amplifier, specifying the building blocks; namely, core amplifier, offset compensation, peak detector (PD), comparators (C_S) and shift-register (SR). The core amplifier is detailed in Fig. 4.52, including the four stages (A_{1-4}), the gain control circuit, the output buffer and input resistances to achieve 50 Ω matching.

Fig. 4.52 Detail of amplifier core area

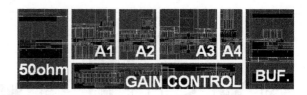

Fig. 4.53 Measured S_{21} parameter for PGA on PCB after de-embedding

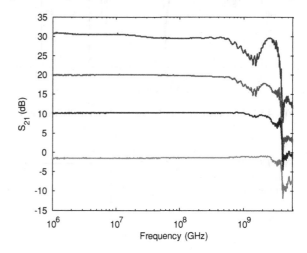

4.5.3 PCB Characterization

The first prototype was packaged in QFN24 and mounted on PCB to easily perform the experimental measurements. Two different PCBs were fabricated, including the input common-mode voltage by bias-tee or by a passive network implemented on the PCB. Furthermore, 50 Ω matching resistors and coupling capacitors at the output were added. The supply and biasing signals were filtered by typical passive networks. The 3-bit digital word was generated from the supply by a set of switches mounted on the PCB.

Frequency-domain measurements were made using an R&SZV6 vector network analyzer. Owing to the required 50 Ω matching, S parameters (see Sect. A.1.1) are frequently employed for high-frequency measurements. Figure 4.53 shows the obtained S_{21} after de-embedding for the gain states separated by 12 dB gain step ($n = 1, 3, 5$, and 7). The S_{11} and S_{22} parameters for these gain states are shown in Fig. 4.54.

Figure 4.53 clearly shows the four different gain states, as the S_{21} parameter is closely related to the gain of the amplifier. The result illustrated in Fig. 4.54 for S_{11} and S_{22} does not depend on the gain state, because the implemented gain variation technique does not affect the input and output nodes. The value of these two latter parameters was found to be excellent for low frequency (<-20 dB), and remained below -10 dB for S_{11} over the measured frequency range; however, such a value exceeded at 3 GHz for S_{22}.

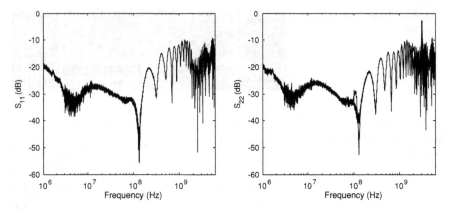

Fig. 4.54 Measured S_{11} and S_{22} parameters for PGA on PCB

Fig. 4.55 Measured gain
compared to expected
distribution

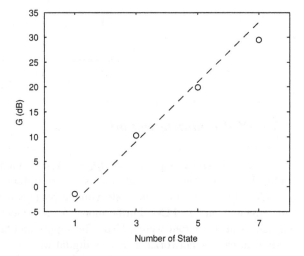

High-frequency behavior of PGA prototype and de-embedding circuit is
dominated by a peak at the frequency of interest. In addition, for the two highest
gain states, de-embedding technique does not cancel this effect. Therefore, we
were not able to determine the high-frequency behavior of the post-amplifier and
time-domain measurements could not be performed on the design.

Fortunately, low-frequency performance could be tested. The measured power
consumption of the chip, including post-amplifier and de-embedding circuit, was
106.2 mW, validating the simulated value, where 54 mW corresponded to
post-amplifier. The voltage gain A_V, calculated from S parameters (A.11), as a
function of the digital control word, depicted in Fig. 4.55, reveals a clear
discrepancy with the expected distribution, especially for the highest gain state.

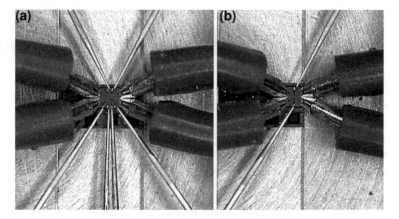

Fig. 4.56 **a** PGA prototype and **b** driver with measurement probes

Fig. 4.57 Measured S_{21}
parameter for PGA
implementation after
de-embedding

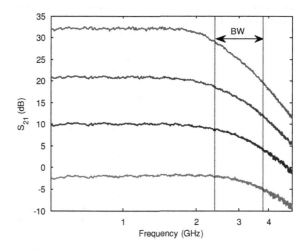

4.5.4 On-Wafer Characterization

To test the prototypes directly without packaging to avoid parasitic effects, an on-wafer probe station is required. This equipment provides a direct access to the output PADs of the prototype, but only if the pin-out shows a strict structure.

The input and output for RF signals, formed by ground-signal-ground (GSG) for single-ended and GSGSG for differential signal, must be located oppositely with fixed spacing. In addition, the required DC signals can be included, but the maximum number is very limited. In this case, the maximum number with individual probes (and one double) is two DC signals from "north" and four from "south." To relax this limitation, a multiple DC probe can be used. However, the more the number of probes, the more complicated will be the alignment and the higher will be the chip area.

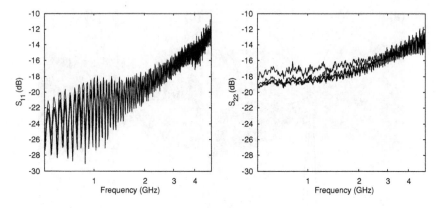

Fig. 4.58 Measured S_{11} and S_{22} parameters for PGA implementation

Fig. 4.59 Measured gain distribution for PGA implementation

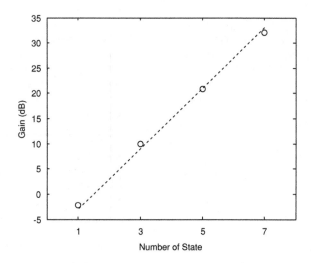

Furthermore, calibration and de-embedding must be performed with further accuracy. Thus, parasitic effects, related to measurement equipment and input and output on-chip signal path, respectively, can be taken into account, and finally, be compensated to attain an accurate characterization. Measurements were performed at room temperature and nominal conditions, including a recent calibration and with no isolation for wafer.

4.5.4.1 Programmable Gain Amplifier Results

Figure 4.56 shows the PGA prototype and de-embedding circuit ready to be tested. The active area is $150 \times 50 \ \mu m^2$ and the chip size, including PAD structure, is 0.9×1

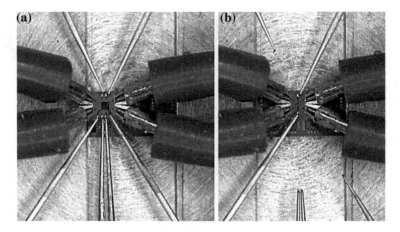

Fig. 4.60 **a** AGC prototype and **b** driver with measurement probes

Fig. 4.61 Measured S_{21} parameter for AGC implementation after de-embedding

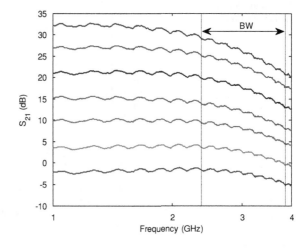

mm^2. PGA and driver implementations consumed 72 and 42 mA respectively, leading to a 54 mW power consumption of the PGA core at a single supply voltage of 1.8 V.

Frequency-domain measurements were made using an R&SZVB8 vector network analyzer. The frequency range was limited from 500 MHz to 5 GHz to achieve a good calibration, while the gain and bandwidth of the post-amplifier could be properly characterized. In this case, the frequency responses with no peak were obtained for prototype and de-embedding circuit; thus, previous undesirable high-frequency effect might have come from PCB.

Figure 4.57 shows the obtained S_{21} after de-embedding for the same gain states as the previous case, because the implementation was equivalent. In this case, the gain and the bandwidth of the four different gain states were characterized. As shown in Fig. 4.58, the S_{11} and S_{22} parameters remained below -10 dB for the

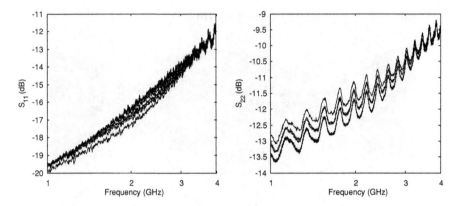

Fig. 4.62 Measured S_{11} and S_{22} parameters for AGC implementation

Fig. 4.63 Measured gain
distribution for AGC
implementation

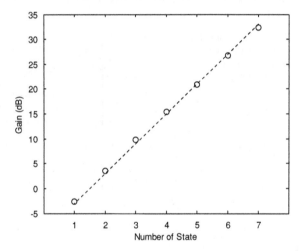

whole frequency range tested, and, as the previous case, did not depend significantly on the gain state.

The voltage gain A_V as a function of the digital control word is depicted in Fig. 4.59, revealing a maximum deviation of 1.0 dB for $n = 7$. A higher accuracy between the expected distribution and measured data was obtained owing to a more careful layout design and other effects avoided by on-wafer technique. The measured -3 dB bandwidth was above 2.3 GHz ($n = 7$) and below 3.7 GHz ($n = 1$) over a 33 to -3 dB linear-in-dB gain range. When compared with the simulation results shown in Table 4.3, the measured bandwidth matches with the values obtained using no negative capacitances technique.

Noise performance was also tested. RMS output noise for the highest gain state was below 5.6 mV. From this result and highest gain (32 dB), an input referred RMS noise of 0.14 mV$_{\text{p-p}}$ could be derived.

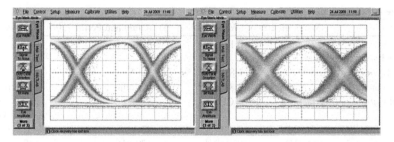

Fig. 4.64 **a** 2.5 Gb/s and **b** 3.125 Gb/s measured eye diagrams NRZ data with PRBS of $2^{31}-1$ at 250 mV output amplitude peak-to-peak for highest gain state

Fig. 4.65 Measured bit rate versus input signal level

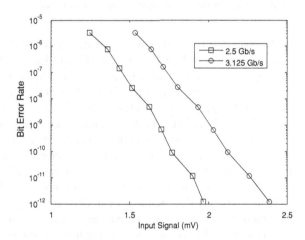

4.5.4.2 AGC Post-Amplifier Results

Measurement process carried out for PGA was repeated for the complete AGC implementation, as shown in Fig. 4.60. In this case, the active area is $300 \times 300 \ \mu m^2$ and the chip size, including PAD structure, is $0.9 \times 1 \ mm^2$. AGC and driver implementations consumed 74 and 42 mA, respectively, leading to a 58 mW power consumption of the AGC core at a single supply voltage of 1.8 V.

The frequency range employed in this case (1–4 GHz) was even more limited than that used to characterize the PGA core, but it was sufficient to extract the gain and bandwidth. Figure 4.61 shows the obtained S_{21} after de-embedding for all gain states of the post-amplifier, while Fig. 4.62 illustrates the results for S_{11} and S_{22}.

In this case, all different gain states (seven) are reflected in Fig. 4.61. The S_{11} parameter remained below -10 dB for the whole frequency range tested, while S_{22} was degraded up to -9 dB for high frequencies. As usual, both were quite independent from the gain state. In spite of the ripple detected in this case, the gain and

Table 4.9 Comparison of several post-amplifiers

Design	(Wu et al. 2004)	(Wang et al. 2005)	(Crain and Perrot 2006)	This work
CMOS technology	0.18 μm	0.18 μm	0.18 μm	0.18 μm
Supply voltage	1.8 V	1.8 V	1.8 V	1.8 V
Active area	0.7 mm^2	0.83 mm^2	0.5 mm^2	0.1 mm^2
Bandwidth	2 GHz	4 GHz	5 GHz	2.3–3.7 GHz
Gain	−16–34	0–20	42	−2.1–32
RMS input noise	640 μV[a]	930 μV[a]	179 μV[b]	140 μV
Power dissipation	40 mW	55 mW	113 mW	58 mW
FOM	0.1	0.1	0.5	2.8

[a] Calculated from reported sensitivity
[b] Calculated from the lowest input amplitude considered

the bandwidth for each state was evaluated quite accurately, taking into account an average value from 1 to 1.5 GHz for the mid-band gain.

The voltage gain A_V as a function of the digital control word is depicted in Fig. 4.63, again revealing a maximum deviation of 1.0 dB for $n = 7$. The measured −3 dB bandwidth was above 2.3 GHz ($n = 7$) and below 3.8 GHz ($n = 1$) over a 33 to −3 dB linear-in-dB gain range. These results confirm the reliability of the gain distribution and the high-frequency behavior of the post-amplifier core.

For the time-domain measurements, a setup was used consisting of Agilent E8257C signal generators, DCA-J 86117A electrical scope modules, an Anritsu MP 1775A pulse-pattern generator, and a proprietary 40 Gb/s multiplexer.

Figure 4.64 depicts the output eye diagrams at 2.5 and 3.125 Gb/s NRZ data with PRBS 2^{31}–1 at 250 mV output amplitude peak-to-peak for highest gain state. The measured BER vs. input signal amplitude is shown in Fig. 4.65. These results demonstrate a sensitivity below 2 mV$_{p-p}$ at 2.5 Gb/s, according to the input referred RMS noise, and 2.5 mV$_{p-p}$ at 3.125 Gb/s due to ISI and jitter penalty. The measured dynamic range for 3.125 Gb/s was 50 dB, from 2.5 mV$_{p-p}$, indicating a BER of 10^{-12}, to 800 mV$_{p-p}$, avoiding saturation effects. The settling time of the AGC loop was measured for a gain increase and a gain reduction of 24 dB. In both the cases, it remained below 1 μs.

4.6 Conclusions

This chapter has covered an in-depth design analysis of the second electrical circuit of the optical receiver: the post-amplifier. The goal of this building block consists of boosting the voltage provided by the TIA to logical levels, adequate for the subsequent clock and data recovery (CDR) and decision circuits. Its main performances can be summarized as follows: gain, which must be high; speed, quantified by the bit rate of transmission; the input-referred RMS noise, being a small penalty to the sensitivity targeted by the TIA; and the input dynamic range, from a lowest to highest value defined by noise performance and overload effects, respectively. The bandwidth of the post-amplifier should be approximately equal to the required bit rate so as not to degrade the optimized bandwidth of the TIA.

The requirements of high gain and broad bandwidth lead to a high GBW, which is limited by every technology for single-stage designs. This issue may be overcome by a multi-stage design, as the total gain is the product of each single gain and the total bandwidth is not reduced in the same factor. Although the theoretical optimum number of stages is quite high, power and area consumptions significantly limit the number of stages. However, the attained GBW extension factor represents a critical improvement. As the required bandwidth is still more demanding for multi-stage designs, some broadband techniques must be employed to target it, such as inter-stage buffering, inverse scaling, negative capacitances, and zero-pole cancellation. The post-amplifier was designed to target 33 dB gain and 3 GHz bandwidth, although measurements demonstrate bandwidth shrinkage (2.3 GHz), which might indicate that negative capacitances technique was not useful, as the experimental results match with the simulated ones without this technique.

In addition to the high-performance core amplifier, the complete AGC consists of two feedback loops: AGC and DC offset compensation loop. In feedback VGA topologies, a constant settling time is achieved by a linear-in-dB gain distribution. As constant output amplitude is not a strict requirement, a discrete gain distribution is explored, leading to a more robust digital control. The linear-in-dB gain distribution facilitates the gain control. A wide gain range (>30 dB) with a tolerant step size (6 dB) leads to a high number of gain states, and hence, a sequential gain variation shows long settling time. A double-step variation diagram is proposed to reduce the settling time below 1 μs. A DC offset compensation loop is mandatory to compensate PVT variations due to the high gain of the post-amplifier. The low-frequency cut-off caused by this loop must be sufficient low to avoid a significant sensitivity penalty. In addition, the output offset voltage might produce a bad function of the AGC loop if it is based on the detection of the output peak value.

The proposed design consists of three cascaded gain stages and a fourth stage implementing the inter-stage buffering technique. In addition, inverse scaling between the first and second stage, negative capacitances in the third stage, and zero-pole cancellation in the fourth stage are applied. Gain stage is based on the common-mode feedforward pseudo-differential pair introduced by Snelgrove. Our proposal is suitable for low-voltage operation as a maximum of only two transistors are connected between supply and ground. The gain is modified owing to a variable load resistor, implemented by an NMOS array to attain the linear-in-dB gain distribution by a thermometer code control. The double-step gain control is formed by a peak detector, four comparators, and a double-direction shift register, while the offset is compensated by controlling a piece of a current source of first gain stage using a basic RC integrator, implemented by transistors to reduce area consumption and an inverting amplifier.

Three prototypes have been integrated in 0.18 μm standard CMOS technology. The first one includes the PGA core, and it was packaged to test on PCB. The frequency behavior shows an undesirable effect that avoids proper characterization. Measurements on the second prototype that also includes the PGA to test on-wafer validate the expected gain distribution with a maximum gain error of 1 dB. Although the measured bandwidth shows shrinkage, when compared with the simulated value, the post-amplifier is able to operate at multi-gigabit transmission speed. The third prototype that includes the complete AGC implementation to be tested on-wafer corroborates the frequency response of the PGA core. In addition, time-domain measurements have also been performed in this case. Measured eye diagrams show a proper operation of up to 3.125 Gb/s, with a sensitivity below 2.5 mV$_{p-p}$. The sensitivity is improved to 2 mV$_{p-p}$ for 2.5 Gb/s. In conclusion, based on a low-cost technology, the proposed post-amplifier complies with the objective of targeting the 10GBase-LX4 Ethernet standard (IEEE std 2003).

To make a fairer comparison, including speed, noise performance, and power and area consumption, a figure of merit (FOM) is defined as:

$$FOM = \frac{Bandwidth}{RMS\,Input\,Noise \cdot Power \cdot Area} \left[\frac{GHz}{\mu V \cdot W \cdot mm^2} \right] \qquad (4.41)$$

Table 4.9 shows the main performances of the proposed design, when compared with those of several previous realizations in 0.18 μm CMOS technology: Wu et al. (2004), Wanget al. (2005), and Crain and Perrot (2006) describe a VGA, an AGC, and a limiting amplifier, respectively. The presented circuit provides the best figure of merit defined by (4.41) owing to significantly lower area consumption, a similar result in terms of power dissipation-bandwidth ratio and a good result in terms of RMS input noise.

Therefore, when compared with the state of the art, the proposed AGC post-amplifier shows several advantages. First, an inductorless design is implemented, thus reducing the active area, and hence, the overall cost. Second, a good result in terms of noise performance is validated with the reported measurement results, improving the sensitivity of the complete optical receiver. Finally, the 34-dB linear-in-dB gain distribution provides a wide input dynamic range and the targeted bandwidth is suitable for multi-gigabit optical transmission.

References

Alegre JP, Celma S, Calvo B, García del Pozo JM (2008) Design of a novel envelope detector for fast-settling circuits. IEEE Trans Instrum Meas 57:4–9

Alegre JP, Celma S, Calvo B (2011) Automatic gain control: techniques and architectures for RF receivers. Analog circuits and signal processing. Springer, Berlin

Atef M, Swoboda R, Zimmermann H (2008) An automatic gain control front-end optical receiver for multi-level data transmission. In: Proceedings of the 26th Norchip conference, pp 57–60

Aznar F, Celma S, Calvo B, Digón D (2008a) Inductorless AGC amplifier for 10GBase-LX4 ethernet in 0.18 μm CMOS. Electron Lett 44(6):409–410

Aznar F, Celma S, Calvo B, Digón D (2008) A fully integrated inductorless agc amplifier for optical gigabit ethernet in 0.18 µm CMOS. In: Proceedings of the 2008 IEEE international symposium on industrial electronics, pp 1662–1667

Aznar F, Celma S, Calvo B, Aldea C (2009) A 0.18 µm CMOS inductorless AGC amplifier with 50 dB input dynamic range for 10GBase-LX4 ethernet, VLSI circuits and systems IV, Proceedings of SPIE, vol 7363, 73630T-1

Aznar F, Celma S, Calvo B (2011) A 0.18 µm CMOS linear-in-dB AGC post-amplifier for optical communications. Microelectron Reliab 51:959–964

Calvo B, Celma S, Sanz MT, Alegre JP, Aznar F (2008) Low-voltage linearly tunable CMOS transconductor with common-mode feedforward. IEEE Trans Circuits Syst I 55(3):715–721

Crain EA, Perrot MH (2006) A 3.125 Gb/s limit amplifier in CMOS with 42 dB gain and 1 µs offset compensation. IEEE J Solid-State Circuits 41(2):443–451

Gajski DD (1997) Principles of digital design. Prentice Hall, Englewood Cliffs

García del Pozo JM (2010) Design of CMOS analog front-ends for broadband optical receiver. PhD thesis, University of Zaragoza, Spain

Green D (1983) Global stability analysis of automatic gain control circuits. IEEE Trans Circuits Syst 30(2):78–83

Hermans C, Steyaert M (2005) A 3.5 Gbit/s post-amplifier in 0.18 µm CMOS. In: IEEE European Solid State Circuits Conference 431–434

Hermans C, Steyaert M (2007) Broadband opto-electrical receivers in standard CMOS, analog circuits and signal processing. Springer, Berlin

Israelsohn J (2002) Gain control. EDN 38–46

Jindal RP (1987) Gigahertz-band high-gain low-noise AGC amplifiers in fine-line NMOS. IEEE J Solid-State Circuits SC-22(4):512–521

Khoury JM (1998) On the design of constant settling time AGC circuits. IEEE Trans Circuits Syst II, Analog Digital Signal Process 45(3):283–294

Lin C-W, Liu Y-Z, Hsu KYJ, (2004) A low distortion and fast settling time automatic gain control amplifier in CMOS technology. In: Proceedings of the 2004 IEEE international symposium on circuits and systems, vol. I, pp 541–544

Maxim Integrated Products (2008) NRZ bandwidth—LF cutoff and baseline wander, application note HFAN-09.0.4, rev. 1

Miller JM (1920) Dependence of the input impedance of a three-electrode vacuum tube upon the load in the plate circuit. Scientific Papers of the Bureau of Standards, vol 15, No 351, pp 367–385 Available on-line at: http://web.mit.edu/klund/www/papers/jmiller.pdf

Muller P, Leblebici Y (2007) CMOS multichannel single-chip receivers for multi-gigabit optical data communications analog circuits and signal processing. Springer, Berlin

McCreary JL (1981) Matching properties, and voltage and temperature dependence on MOS capacitors. IEEE J Solid-State Circuits SC-16(6):608–616

Park S-B, Wilson JE, Ismail M (2006) Peak detectors for multistandard wireless receivers. IEEE Circuits Devices Mag 22(6):6–9

Razavi B (2001) Design of high-speed circuits for optical communication systems. IEEE Custom Integr Circuits Conf 315–322

Razavi B (2008) Fundamentals of microelectronics. Wiley, Hoboken

Säckinger E, Fischer WC (2000) A 3-GHz 32-dB CMOS limiting amplifier for SONET OC-48 reveivers. IEEE J Solid-State Circuits 35(12):1884–1888

Säckinger E (2005) Broadband circuits for optical fiber communication. Wiley, Hoboken

Sanz MT, Celma S, Calvo B (2005) High linear digitally programmable gain amplifier. In: Proceedings of the 2005 IEEE international symposium on circuits and systems, vol 1, pp 208–211

Schneider K, Zimmermann H (2006) Highly sensitive optical receivers, Springer series in advanced microelectronics. Springer, Berlin

Snelgrove WM, Shoval A (1992) A balanced 0.9-um CMOS transconductance-C filter tunable over the VHF range. IEEE J Solid-State Circuits 27(3):314–322

IEEE Std. 802.3af-2003

Tsai C-M, Chen W-T (2007) A 40 mW 3.5 kΩ 3 Gb/s CMOS differential transimpedance
    amplifier using negative-impedance compensation. IEEE Int Solid-State Circuits Conf,
    pp 52–53, 586
Wang I-H, Chen W-S, Liu S-I (2005) A 5 Gbps CMOS automatic gain control amplifier for
    10GBase-LX. In: Proceedings of the 2005 IEEE Asian solid-state circuits conference,
    pp 169–172
Wu C, Liu C, Liu S (2004) A 2 GHz CMOS variable-gain amplifier with 50 dB linear-in-
    magnitude controlled gain range for 10GBase-LX4 ethernet. IEEE Int Solid-State Circuits
    Conf 1:484–541

Chapter 5
POF Receiver

Traditionally, optical fiber communications have been exploited for shared long-haul communication links. In such cases, the high cost of transmitter and receiver, fabricated using GaAs or InP technology to achieve the standard requirements, is compensated by the huge number of users. However, the ever-increasing amounts of data transmitted over short-distances (50 m) mandate the cost efficiency of the system.

For such an application, the economic viability requires the use of low-cost technologies for both microelectronic and optical components. The electronic front-end can be fabricated using a standard CMOS technology. For the optical channel, polymethyl-methacrylate or 'plastic' optical fiber (POF) is a cost-effective choice (Polishuk 2006), what has encouraged its development.[1]

In this chapter, the main limitations of the low-cost optical channel, such as attenuation and bandwidth–length product, are introduced. The selected technique to compensate the low speed based on equalization is then analyzed. Thus, a fully integrated front-end receiver implemented in 0.18 μm CMOS technology (Aznar et al. 2010), using the previously presented TIA and post-amplifier plus an adaptive equalizer and a photodiode monitor, is proposed for fiber-to-the-home (FTTH) applications (Koonen et al. 2011) at 1.25 Gb/s transmission speed on POF.

5.1 Plastic Optical Fiber

Fundamentals of optical fibers were introduced in Sect. 2.2. However, the speed limitations caused by the optical channel were omitted, mainly because it was supposed that the speed bottleneck was caused by the opto-electronic conversion in the receiver. For POF, the limitations derived from the fiber itself, such as attenuation and bandwidth, become critical (Ziemann et al. 2008).

---

[1] POF-PLUS. http://www.ict-pof-plus.eu

F. Aznar et al., *CMOS Receiver Front-ends for Gigabit Short-Range Optical Communications*, Analog Circuits and Signal Processing, DOI: 10.1007/978-1-4614-3464-1_5, © Springer Science+Business Media New York 2013

POFs are usually formed by poly-methyl methacrylate (PMMA). The complex structure of the organic compound leads to a relatively high attenuation. Typical loss at 650 nm is 0.14 dB/m, which is suitable only for short-haul transmission.

Due to the typical loss of POF, the distance from emitter to receiver could be higher than 50 m; however, the bandwidth of the fiber is a more restrictive limitation (Sundermeyer et al. 2009). Let us introduce the concept of optical fiber bandwidth, and then analyze the bandwidth–length dependency of POF.

5.1.1 Optical Fiber Bandwidth

There are two different concepts of bandwidth that are related to optical fibers (Säckinger 2005). Each one is derived from a non-ideal effect of the optical fiber, namely, attenuation and dispersion. Both the effects are illustrated in Fig. 5.1 for a single-mode glass optical fiber, which is used as an example to introduce the concepts of bandwidth and to show the potential speed of optical fibers.

Minimal attenuation is attained at 1.55 μm window. The first concept of bandwidth is defined as the frequency range where attenuation is close to the minimal. This bandwidth is also known as low-loss window (around 100 nm) and leads to a value higher than 10 THz. Therefore, an optical signal modulated under a non-return-to-zero (NRZ) format at a multi-terabit data rate can be transmitted, as its entire frequency spectrum would be within the low-loss window.

Nevertheless, the dispersion effect limits the highest data rate which could be transmitted by several orders of magnitude. From Fig. 5.1, dispersion is around 17 ps/(nm·km) at the lowest attenuation. This means that the pulses will spread 17 ps for a 1 km piece of optical fiber with a 1 nm linewidth. Thus, the bandwidth—derived from twice the value of the dispersion to avoid overlapping—is around 30 GHz. Then, the transmission is limited to multi-gigabit speed.

As shown in Fig. 5.1, for a standard single-mode GOF, a wavelength of 1.55 μm offers minimal loss, whereas a wavelength of 1.3 μm gives minimal dispersion. This dilemma was solved by introducing dispersion-shifted fibers (Budin 1989), which compensated for the dispersion at 1.55 μm while minimal losses were preserved. Thus, the bit rate–distance product of GOF was optimized.

To exploit the entire low-loss window of an optical fiber, dense wavelength-division multiplexing can be employed (Park et al. 2004). Thus, 10 Gb/s signals can be transmitted at more than 100 different wavelengths, attaining a multi-terabit data rate over a single fiber.

5.1.2 Bandwidth-Length Dependency

As explained earlier, dispersion leads to a more restrictive speed limitation of the optical channel. In addition, similar to attenuation, this will depend on the length of the fiber, and as a result, a bandwidth-length dependency is derived.

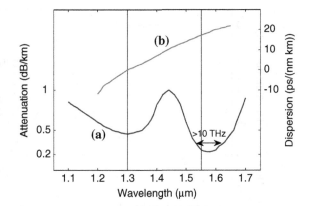

Fig. 5.1 a Attenuation, and **b** dispersion of a typical single-mode GOF

For a typical POF, the length-bandwidth dependency, as shown in Fig. 5.2, is around 40 MHz × 100 m, which makes it difficult to target gigabit communications over a certain length. Furthermore, a constant bandwidth of the fiber joint to the receiver is required to properly process the transmitted signal.

In addition, a core larger than the glass optical fiber leads to a large photodiode that can efficiently detect the transmitted light. Then, the large capacitance of the photodetector directly impacts the design of the receiver, which must show a low input resistance. Therefore, receivers with adaptive gain and equalization are mandatory to compensate for the band-limited frequency response due to the large photodiode required, and to the frequency response of the fiber itself.

5.2 Equalization

The bandwidth of a typical optical fiber can be enhanced by implementing optimized techniques, such as graded-index profile or dispersion-shift fibers (Kawai 2005). However, the fabrication cost increases. Alternatively, equalization can be employed to compensate for the band-limiting effect (Sundermeyer et al. 2009).

A comparison among different equalization techniques implemented in the transmitter or the receiver, and an introduction to adaptive equalization are presented in the following sections.

5.2.1 Equalization Techniques

Equalization can be defined as the frequency-selective-boosting signal processing (Hall and Heck 2009). In telecommunications, it is usually employed to compensate the high cut-off frequency of transmission channels avoiding distortions at high frequency (mainly inter-symbol interference), and then, increasing the data

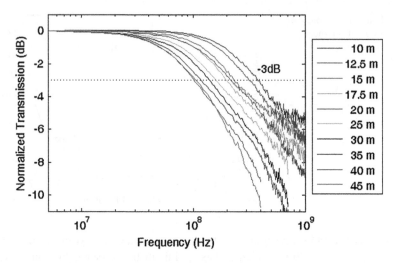

Fig. 5.2 Bandwidth-length dependency

rate and/or improving the bit error ratio (Liu and Lin 2004). Equalizers can be classified depending on diverse criteria: components (active or passive), signal processing (discrete-time and continuous-time), and hardware (analog and digital). In addition, an equalizer can be implemented in the transmitter and/or in the receiver.

The most common implementation of an equalizer in the transmitter is the pre-emphasis technique (Schrader et al. 2005). It consists of boosting the high-frequency spectrum of the data prior to transmission. Another approach consists of weakening the low-frequency spectrum, denoted as de-emphasis (Liu and Lin 2004); however, this worsens the signal-to-noise ratio, and thus, the quality of transmission. Both the implementations show the same drawback: they lack information about the transmission channel. Thus, an adaptive equalizer cannot be implemented, and the equalizer in the transmitter might compensate the frequency response of the channel only for a particular case. Variations due to temperature, aging, or any other effect, require an adaptive equalizer at the receiver (Säckinger 2005).

In the receiver, equalization should be performed before decision circuit; thus, distortions might be corrected and a proper recovery may be achieved. Discrete-time and digital equalizers require additional circuitry (a sampling circuit for discrete-time and, in addition, an analog-to-digital converter for digital equalizers). Furthermore, according to the Nyquist theorem (Nyquist 1928) demonstrated by Shannon (Shannon 1949), the signal must, at least, be sampled at twice the highest frequency, which leads to a very high-speed requirement for a sampling circuit at the cost of high power consumption. Therefore, continuous-time operation is preferable for high-speed applications (Otín 2006).

Fig. 5.3 Typical block diagram of optical receiver front-end

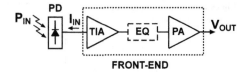

FRONT-END

The typical location of the equalizer in the receiver is between the transimpedance amplifier and the post-amplifier, as shown in Fig. 5.3. The main reasons are that typical equalizer structures are based on voltage amplifiers and the signal level is lower, thereby facilitating signal processing. Active, instead of passive, equalizers are implemented, as inductors requiring a large silicon area are avoided and the input referred noise is optimized (Cheng and Tsai 2010).

In conclusion, a simple continuous-time equalizer has been chosen instead of the more sophisticated digital signal processing-based approaches because it is suitable for low-power high-speed applications and does not require additional circuitry. Hence, it works independent of the clock recovery circuit.

5.2.2 Adaptive Equalization

SI-POF exhibits an inherent frequency limitation, which is not proper for gigabit communications (Ziemann et al. 2008). Furthermore, it depends greatly on the length as shown previously in Fig. 5.2, making adaptive equalizing mandatory for such an application.

Different fiber lengths cause different signal levels due to fiber loss. This is compensated by the gain control implemented in TIA and post-amplifier. Thus, for the simplest equalizer design, a zero adjustment of the frequency response of the equalizer is required while the gain at low frequencies is constant. Figure 5.4 illustrates the typical frequency response behavior of an equalizer with zero adjustment. Because the low-frequency gain and the double-pole frequency location are constant, zero adjustment and gain boosting are coupled. The lower the zero, the higher the peak. The zero location must be controlled by an electrical variable, and the typical choice is a voltage that drives a transistor implementing a variable resistor or capacitor. The requirement in terms of bandwidth is similar to the post-amplifier, as the equalizer might be considered its first stage.

The approach followed to generate the control voltage is based on the power spectrum of the data stream (Couch 2007). The key idea is that if the data have been scrambled, that is, the bit sequence is random and with sufficient length, its power spectrum is described by a predictable mathematical function ($sinc^2$ for NRZ code), as explained in Sect. 2.1.2. Thus, the deviations of the expected frequency spectrum of the output signal can be detected by comparing the power densities of two different frequency ranges. In particular, Fig. 5.5 illustrates the comparison between the power density below a determined frequency (P_L) and

Fig. 5.4 Adaptive equalizer
frequency response

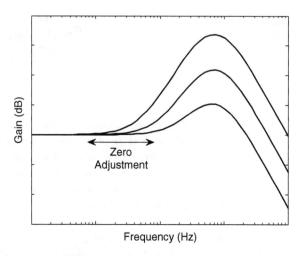

Fig. 5.5 Power spectrum
comparison approach

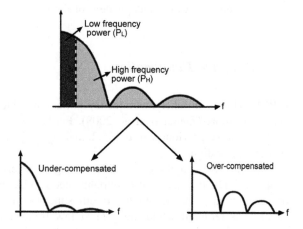

above the same frequency (P_H), both ideally filtered. Then, if the system was under- or over-compensated, the power ratio P_H/P_L would change, detecting the deviation from the ideal situation. A typical block diagram of the adaptive loop based on this approach is shown in Fig. 5.6.

The spectrum power estimator is formed by a low-pass filter (LPF), a high-pass filter (HPF), two rectifiers, a subtractor, and an integrator. According to Fig. 5.5, the LPF and the HPF are ideal and show the same cut-off frequency. The two signal powers are measured by rectifiers. Their difference is driven to the integrator, which provides the control voltage to the equalizer.

It must be noted that the choice of the filters (LPF and HPF) is arbitrary, as the power ratio of any two different filters is known for a particular frequency response, such as the NRZ spectrum. Therefore, this approach can be implemented, for example, with two LPFs with different cut-off frequencies, by replacing the LPF, the HPF, or both by band-pass filters, or even omitting the LPF or HPF. The

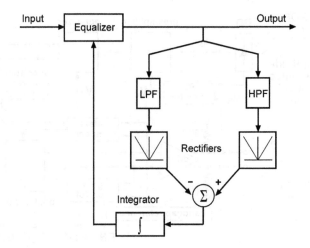

Fig. 5.6 Spectrum power estimator diagram

simplest solution is to use a spectrum power estimator, which compares the entire signal power and the low-pass filtered signal power (Sun et al. 2006).

5.3 Receiver Architecture

Figure 5.7 shows the block diagram of the receiver architecture (Aznar et al. 2010). It is based on the two previously detailed building blocks, the transimpedance amplifier and the AGC post-amplifier, in addition to the mandatory adaptive equalizer. The combination of pre- and post-amplifier with gain control ensures a proper signal quality at the input of the equalizer and at the output of the receiver over a wide input dynamic range. In this case, the denomination of pre- and post-amplifier is adopted to observe that the amplification is done either before or after equalization.

The photodetector (PD) is an off-chip high-speed Si PIN photodiode.[2] The PD monitor indicates the input current; a voltage is read over a photocurrent sampling resistor providing a voltage control V_C. This voltage automatically adjusts the input–output response of the transimpedance preamplifier to the strength of the input signal. Then, an adaptive equalizer boosts the band-limited signal controlled by a spectrum power estimator, targeting the gigabit data rate. The post-amplifier is able to provide a constant level output, thanks to its double control loop of gain and DC offset. Finally, a 50 Ω driver provides the output digital signal.

---

[2] Hamamatsu Photonics. http://www.hamamatsu.com/.

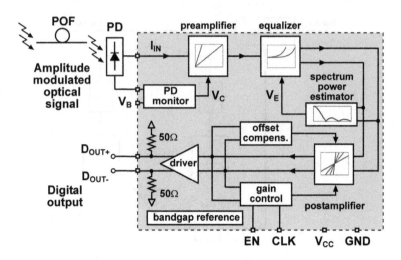

Fig. 5.7 Optical receiver block diagram

5.3.1 Preamplifier with PD Monitor

Although the architecture of the transimpedance amplifier is not modified, some differences must be noted. First, the input resistance must be minimized due to the high photodiode capacitance. In addition, a monitor circuit to modify the transimpedance gain internally was added.

In particular, this prototype has been designed for an off-chip Si PIN photodiode[3] with a large size (active area with diameter of 0.8 mm), a capacitance of 3 pF with 3 V reverse voltage, high responsivity of 0.44 A/W for 650 nm, and high operation frequency of 500 MHz for 50 Ω load resistance. Therefore, the TIA design must be adapted to handle a high capacitance photodiode.

A shunt-feedback structure is selected for the TIA (Aznar et al. 2011); it consists of an inverting amplifier, formed by three stages ($N_1 - P_1$, $N_2 - P_2$ and $N_3 - R$), and a shunt-feedback resistance R_F, as shown in Fig. 5.8. In addition, it includes two transistors N_4-N_5 to vary the transimpedance gain. The TIA design is detailed in Chap. 3. A low-cost fully integrated receiver mandates the 0.18 μm CMOS technology, whereas the high capacitance of the photodiode leads to the selection of a 1 kΩ feedback resistor to adapt the bandwidth to gigabit transmission. No more modifications are required for the previously presented TIA parameters. Although the sensitivity of the receiver is degraded when compared with higher shunt-feedback resistance, the attenuation of short-range transmission is low enough to enable error-free recovery of the data signal.

---

[3] Hamamatsu Photonics, Si PIN Photodiode

Fig. 5.8 Transimpedance
amplifier schematic

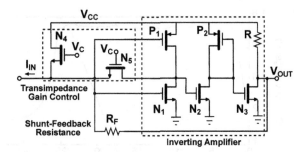

Fig. 5.9 Photodiode monitor
schematic

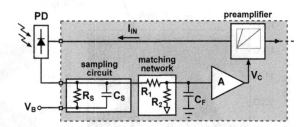

In order to provide a wide input dynamic range, a photodiode monitor was added to modify the transimpedance gain by generating the appropriate control voltage depending on the input photocurrent. The schematic of the photodiode monitor is illustrated in Fig. 5.9. It basically consists of a sampling circuit formed by the resistor R_S and the capacitor C_S, a matching network formed by two resistors R_1 and R_2, a filtering capacitor C_F and an amplifier of gain A and output common-mode voltage of V_{CM}. This circuit provides the control voltage V_C according to (5.1), while the photodiode is biased reversely through the external negative voltage V_B.

$$V_C = V_{CM} + \frac{R_S A}{1 + R_1/R_2} \frac{I_{IN}}{2} \tag{5.1}$$

Figure 5.10 illustrates the time response of the photodiode monitor. The simulation shows the behavior of the system for a variation from high (1.5 mA) to low input signal (15 µA) at 25 µs. In the top picture, the control voltage V_C is shown, demonstrating the adaptability to both corner possibilities. In addition, the eye diagrams during stationary conditions for both cases are shown. Table 5.1 summarizes the main design values of the photodiode monitor.

In the future, low-cost solutions mandate the full integration of the optical receiver and the photodiode in a standard CMOS process (Muller and Leblebici 2007). Thus, cost optimization could be achieved, sacrificing responsivity: typically, $R = 0.05$ A/W (Tavernier and Steyaert 2008) for a photodiode in submicron CMOS technologies. Despite further penalty on sensitivity, there are ever-increasing applications in short-distance networks for the full integrated receiver in CMOS (Hermans and Steyaert 2007).

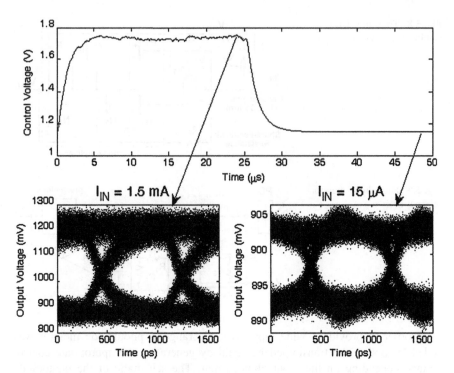

Fig. 5.10 Functionality of the photodiode monitor. Generated control voltage (*top*) for *high* (1.5 mA from 0 to 25 µs) and *low* (15 µA from 25 to 50 µs) peak to peak input current and both eye diagrams (*bottom*) at TIA output during stationary conditions

Table 5.1 Design parameters of photodiode monitor

Instance	Value
R_S	480 Ω
C_S	10 pF
R_1	400 kΩ
R_2	80 kΩ
C_F	20 pF
A	20 dB

5.3.2 Adaptive Equalizer Implementation

Adaptive equalization is based on the simplest spectrum power comparison topology, that is, by comparing the entire signal with the low-pass filtered signal, as explained in Sect. 5.2.2. Therefore, the adaptive equalizer consists of a voltage-controlled equalizer and the spectrum power estimator, formed by a low-pass filter (LPF) and a power error detector, as shown in Fig. 5.11.

Thus, the total (P_{TOT}) and the low-pass filtered (P_{LPF}) signal power, shown in Fig. 5.12, are compared and a particular relationship (5.2) between them is only

Fig. 5.11 Selected adaptive
equalizer loop

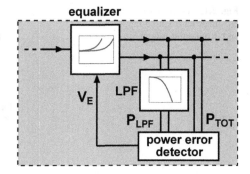

Fig. 5.12 Entire and low-
pass filtered signal power

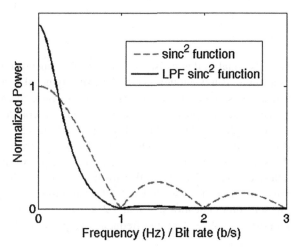

valid for the desired situation, as the frequency spectrum of the data signal is
known. The power error detector provides a control voltage V_E depending on the
power ratio. In this case, the control voltage is denominated as the error signal, as
no signal is generated for the optimum case.

$$\frac{P_{TOT}}{P_{LPF}} = const. \tag{5.2}$$

5.3.2.1 Voltage-Controlled Equalizer

The proposed equalizer, as shown in Fig. 5.13, consists of two stages: a common
source and a degenerated differential amplifier (Kuo and Leuciuc 2001). The first
stage formed by the transistor N_1 and resistor R_1 is used to provide a common-
mode voltage of 1.2 V at the input of the differential amplifier, formed by the
$N_2 - N_3$ pair and the resistors R_2 and R_3. In order to generate a differential output,
a passive low-pass filter (R_0, C_0) is implemented between the inputs of the differential amplifier, which creates a low-frequency cut-off below 100 kHz. This

Fig. 5.13 Proposed equalizer with single input and differential output

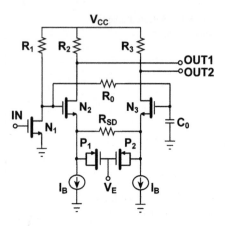

value is equal to that caused by the offset compensation of the post-amplifier, and as a whole, is low enough to avoid sensitivity penalty due to low cut-off frequency. Thus, it is suitable for a wide range of communication standards.

The equalizer provides an additional zero to compensate the roll-off frequency caused by the SI-POF and PD. The zero location is adjusted by the error voltage V_E driving a PMOS varactor (Otín 2006) formed by transistors $P_1 - P_2$, in parallel to the source degenerated resistor R_{SD}. The frequency position of the zero can be adjusted from 125 to 385 MHz for a V_E variation from 1.5 to 0.6 V; a sufficient range to cover the possible bandwidth variations and to fulfill the specifications of gigabit communications. Table 5.2 summarizes the main design values of the proposed equalizer, and the frequency response is illustrated in Fig. 5.14.

5.3.2.2 Low-Pass Filter

The architecture of the low-pass filter, shown in Fig. 5.15, is based on a fully differential source-degenerated amplifier (Kuo and Leuciuc 2001) formed by transistors $N_1 - N_2$ and the MOS resistor N_3. To achieve sufficient gain with only one stage, a positive feedback is included through PMOS transistors $P_2 - P_3$ (Lin et al. 2004). The cut-off frequency is defined by capacitance C_{LPF} and the trans-conductance, which is externally controlled by voltage V_{LPF}. To facilitate comparison, the LPF is designed to compensate the power losses at high frequency by amplifying the signal level, as illustrated in Fig. 5.12. Thus, (5.2) can be particularly written as (5.3); consequently, the low-frequency gain and the cut-off frequency of the LPF are coupled. In addition, the input and output common-mode voltages of the LPF are equal.

$$P_{TOT} = P_{LPF} \tag{5.3}$$

In particular, for an ideal brick-wall LPF, the relationship between the DC gain A and the bandwidth BW derived from (5.3) can be written as (5.4), where R_b is the

Table 5.2 Design parameters of proposed equalizer

Instance	Width/value	Length
N_1	10 μm	180 nm
$N_2 - N_3$	60 μm	180 nm
$P_1 - P_2$	30 μm	30 μm
I_B	3 mA	
R_1	450 Ω	
$R_2 - R_3$	300 Ω	
R_{SD}	500 Ω	
R_0	500 kΩ	
C_0	5.4 pF	

Fig. 5.14 Simulated frequency response of equalizer depending on error voltage

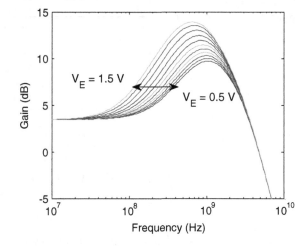

Fig. 5.15 Low-pass filter schematic

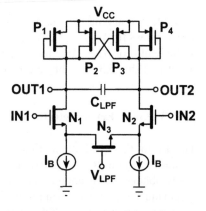

bit rate and sinc(x) function is the normalized frequency spectrum of NRZ signal, defined in (2.6). For a real filter with frequency response H(x), the relationship can be written as (5.5). Some reference values are summarized in Table 5.3.

Table 5.3 Gain and bandwidth combinations depending on frequency response of LPF

LPF	Ideal	First-order	Butterworth second-order
H(x) =	$x \leq BW \Rightarrow A$ $x > BW \Rightarrow 0$	$\dfrac{A \times BW}{x + BW}$	$\dfrac{A \times BW^2}{x^2 + \frac{x \times BW}{\sqrt{2}} + BW^2}$
BW/R_b	Gain A (dB)		
0.01	16.99	15.12	16.53
0.05	10.02	8.50	9.57
0.1	7.04	5.89	6.65
0.15	5.34	4.56	5.01

Table 5.4 Design parameters of low-pass filter

Instance	Width/value	Length
$N_1 - N_2$	60 μm	0.5 μm
N_3	30 μm	0.5 μm
$P_1 - P_4$	28 μm	0.5 μm
$P_2 - P_3$	3.54 μm	0.5 μm
I_B	300 μA	
C_{LPF}	1.3 pF	

$$\int_0^\infty \text{sinc}^2(x)dx = A^2 \int_0^{BW/R_b} \text{sinc}^2(x)dx \tag{5.4}$$

$$\int_0^\infty \text{sinc}^2(x)dx = \int_0^\infty \text{sinc}^2(x)\text{H}^2(x)dx \tag{5.5}$$

Table 5.4 summarizes the main design values of the designed low-pass filter, which shows for $V_{LPF} = 1.2$ V, as presented in Fig. 5.16, a DC gain of 8.6 dB and a bandwidth of 62.3 MHz for a first-order frequency response, according to the result in Table 5.3 for a bit rate R_b of 1.25 Gb/s.

5.3.2.3 Power Error Detector

Thanks to the quadratic current–voltage response of MOS transistors in the saturation region (Razavi 2008), a current proportional to the signal power can be easily obtained. This approximation is only valid for long-channel MOS; hence, a length much superior to the minimum is used. Thus, the implementation of the power error detector based in MOS coupled pair (Zhou and Wah 2006) is shown in Fig. 5.17, where V_T represents the entire voltage signal and V_L is the low-pass filtered voltage signal whose squared values are proportional to the entire and low-pass filtered signal power respectively. Table 5.5 summarizes its main design values.

Fig. 5.16 Simulated
frequency response of LPF

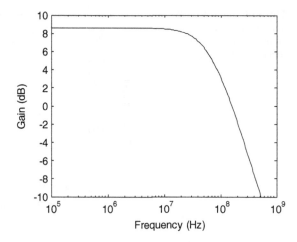

Fig. 5.17 Power error
detector schematic

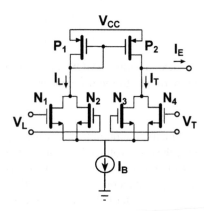

Instance	Width/value	Length
Table 5.5 Design parameters of power error detector		
$N_1 - N_4$	60 μm	0.5 μm
$P_1 - P_2$	26 μm	0.5 μm
I_B	300 μA	

Let us analyze its functionality briefly. Assuming perfectly matched NMOS transistors $N_1 - N_4$, the current I_T is proportional to a quadratic expression including the input common-mode voltage V_{CM}, the common voltage for all NMOS sources V_S, the threshold voltage of NMOS transistors V_{TH}, and the input signal V_T.

$$
\begin{aligned}
I_T &\propto \left[\left(V_{CM} - V_S - V_{TH} + \tfrac{V_T}{2}\right)^2 + \left(V_{CM} - V_S - V_{TH} - \tfrac{V_T}{2}\right)^2 \right] \\
&= 2\left[\left(V_{CM} - V_S - V_{TH}\right)^2 + \left(\tfrac{V_T}{2}\right)^2 \right]
\end{aligned}
\tag{5.6}
$$

Thanks to symmetry, an equivalent expression can be derived for current I_L:

Fig. 5.18 Simulated error voltage generated by amplitude mismatching of input signals

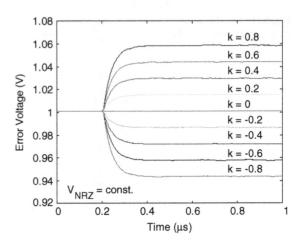

$$I_L \propto 2\left[(V_{CM} - V_S - V_{TH})^2 + \left(\frac{V_L}{2}\right)^2\right] \qquad (5.7)$$

Next, a simple PMOS current mirror $P_1 - P_2$ provides the desired difference I_E, proportional to the differences in signal powers.

$$I_E = I_L - I_T \propto V_L^2 - V_T^2 \qquad (5.8)$$

Finally, the output current is integrated, low-pass filtered, and amplified to provide the error voltage V_E, which adjusts the zero of the equalizer. Simulation results corroborate that the generated error voltage is proportional to the difference in signal powers, as it increases linearly with the mismatching of input signal amplitudes (5.9) introduced after 0.2 µs by k for a NRZ signal V_{NRZ}, as shown in Fig. 5.18.

$$\left.\begin{array}{l} V_L = \sqrt{1+k}\, V_{NRZ} \\ V_T = \sqrt{1-k}\, V_{NRZ} \end{array}\right\} \Rightarrow V_E \propto V_L^2 - V_T^2 \propto k \;\; \text{for} \; |k| \le 1 \qquad (5.9)$$

Figure 5.19 illustrates the time response of the adaptive loop. The simulation shows the behavior of the system for a variation from long (50 m) to short length (10 m) at 25 µs. In the top picture, the error voltage V_E is shown, demonstrating the adaptability to both corner possibilities. In addition, the eye diagrams during stationary conditions for both cases are shown.

5.3.3 Post-Amplifier and Output Driver

The experimental results of the post-amplifier, integrated in 0.18 µm CMOS technology, fulfill the requirements in terms of gain and bandwidth for 1.25 Gb/s

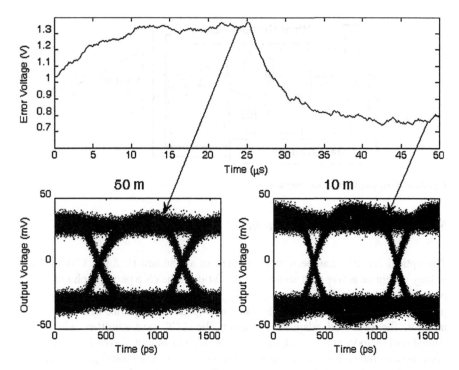

Fig. 5.19 Funtionality of the adaptive loop. Generated error voltage (*top*) for long (0–25 μs) and short (25–50 μs) fiber length and both eye diagrams (*bottom*) during stationary conditions

transmission. Thus, no modifications were required for this building block from that presented in Chap. 4. A block diagram of the structure is illustrated in Fig. 5.20.

Briefly, the post-amplifier (Aznar et al. 2011) consists of a core amplifier formed by four stages including broadband techniques such as inverse scaling, negative capacitances, inter-stage buffering and zero-pole cancellation, and two feedback loops, namely, offset compensation and gain control. Thus, the post-amplifier is implemented as an AGC amplifier and properly processes low and high input signals owing to the output offset minimization and adaptive gain respectively. AGC design is detailed in Chap. 4. Finally, the output driver provides an output signal matched to 50 Ω.

5.4 Experimental Verification

In this section, the experimental verification of the POF receiver is described. First, the receiver implementation and the experimental setup will be presented. Second, the measurement results will be detailed and discussed.

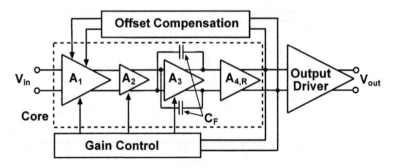

Fig. 5.20 Post-amplifier block diagram

5.4.1 Receiver Implementation

The optical receiver has been implemented in a standard 0.18 μm CMOS technology within an active area about 300 × 300 μm. The chip photograph indicating
pin-out is shown in Fig. 5.21, where T_i pines correspond to test purpose. The
prototype (POFchip-II) is formed by the transimpedance amplifier with photodiode
monitor, the proposed adaptive equalizer, the AGC post-amplifier with offset
compensation loop and an output driver. A second output driver is included to
offer the de-embedding technique. On comparing with the prototype (POFchip-I)
presented in ESSCIRC'10 (Aznar et al. 2010), the supply voltage has been split to
implement on-chip filtering, and additional test pines have been included.

The design is tested on-board for 1 Gb/s and beyond with an NRZ PRBS 2^{31}-1
pattern. Experimental measurements include, as depicted in Fig. 5.22:

- SI-POF Mitsubishi GH.
- Si PIN photodiode Hamamatsu S5972.[4]
- Commercial red laser diode from Thorlabs.[5]
- Digital communications analyzer Agilent 86100C.
- Digital communications analyzer Agilent 86100C.
- Bit error ratio tester Agilent N4906A.

5.4.2 Results

This section has been divided into two parts. First, the simulated frequency
response and eye diagrams with and without equalization will be offered. Next, the
measurement results of the whole receiver including equalization will be analyzed.

---

[4] Hamamatsu Photonics, Si PIN Photodiode
[5] Thorlabs Inc

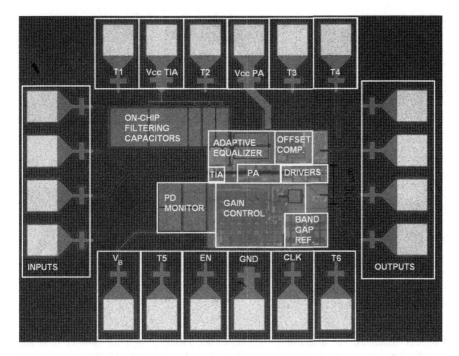

Fig. 5.21 Chip microphotograph

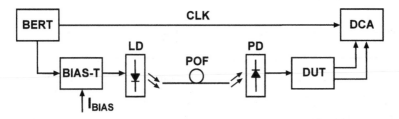

Fig. 5.22 Block diagram of the experimental setup

Accurate models of the POF frequency response for the highest (50 m) and the lowest (10 m) considered distances have been implemented with ideal circuit elements in order to simulate the transmitted eye diagram over POF with and without equalization. The simulation results in time- and frequency-domain including transimpedance amplifier, second stage (as the proposed equalizer shown in Fig. 5.13 or as a single to differential converter omitting PMOS varactor) and post-amplifier are illustrated in Figs. 5.23 and 5.24, respectively.

The presented eye diagrams without equalization shows that the frequency response of the POF is not suitable for 1.25 Gb/s transmission. Fortunately, the proposed adaptive equalizer approach compensates the low-frequency cut-off of POF and provides an open eye diagram for 10 and 50 m, thanks to a bandwidth

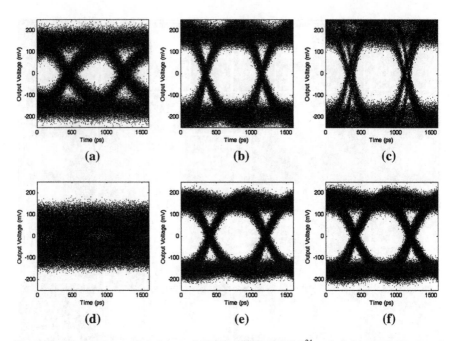

Fig. 5.23 Simulated eye diagrams at 1.25 Gb/s NRZ PRBS $2^{31} - 1$ for (**a–b–c**) 10 m and (**d–e-f**) 50 m and an optical input power nearby sensitivity (-16 and -14 dBm, respectively): (**a–d**) without equalization, (**b–e**) including equalization with constant V_E, and (**c–f**) with V_E generated by adaptive loop

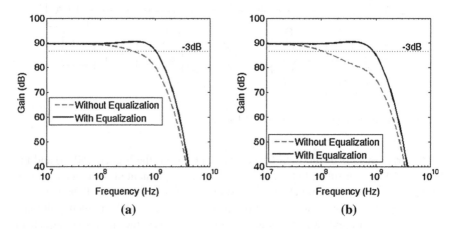

Fig. 5.24 Simulated frequency response without and with equalization for **a** 10 m, and **b** 50 m SI-POF

enhancement from 400 MHz to 1.1 GHz for 10 m and from 100 to 900 MHz for 50 m according to simulated frequency response. The generated error voltage by the adaptive loop shows a ripple, leading to a degradation of the eye diagrams

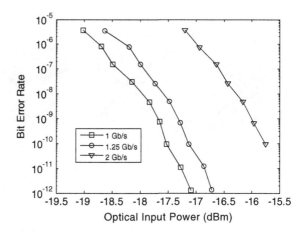

Fig. 5.25 Measured BER versus input optical power for 10 m SI-POF

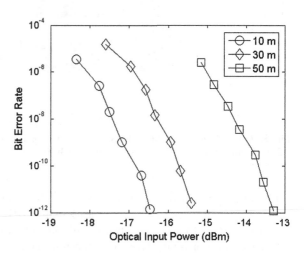

Fig. 5.26 Measured BER versus input optical power depending on SI-POF length

compared with the equalized by a constant error voltage, especially for 10 m length. However, the eye diagrams are still wide open.

Measurements of the integrated prototype POFchip-II in a standard 0.18 μm CMOS technology were performed at room temperature for 10-, 30- and 50-m POF lengths. The total power consumption at a supply voltage of 1.8 V was below 110 mW.

A preliminar characterization were carried out for the prototype POFchip-I. As derived from the measured bit error rate for the 10-m POF length shown in Fig. 5.25, an error-free (BER = 10^{-12}) sensitivity of −16.7 dBm is achieved for 1.25 Gb/s. Furthermore, the receiver is functional up to 2 Gb/s for the same length, but the transmission is not error-free. The definitive experimental verification of the prototype POFchip-II validates an error-free sensitivity of −16.4 dBm. For longer distances, the sensitivity for 1.25 Gb/s is slightly degraded to −15.3 and

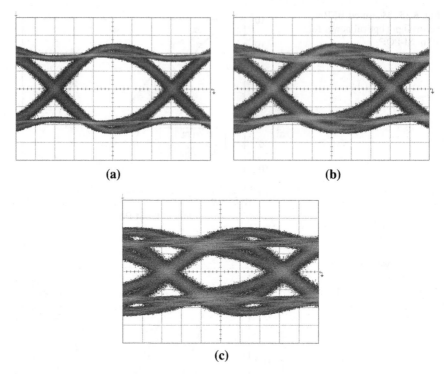

Fig. 5.27 Eye diagrams for NRZ PRBS $2^{31} - 1$ and an optical input power nearby sensitivity at 1.25 Gb/s through **a** 10 m POF length for -16 dBm, **b** 30 m for -15 dBm, and **c** 50 m for -13 dBm

-13.2 dBm for 30 and 50 m respectively, as illustrated in Fig. 5.26. Figure 5.27 shows the measured eye diagrams for a 1.25 Gb/s NRZ signal with an optical input power of -16, -15 and -13 dBm for 10, 30 and 50 m SI-POF respectively.

Nearby the sensitivity level, the transimpedance amplifier works in inactive region (highest transimpedance gain) and the post-amplifier works in a high-gain state, optimizing the input referred noise, and thus, the sensitivity. Owing to the automatic control of the transimpedance and post-amplifier gains, offered by photodiode monitor and AGC loop respectively, the prototype is able to operate with higher input signal levels. The receiver is functional up to 3.5 dBm, leading to an input power dynamic range of 20 dB for 10 m POF. This upper limit of the dynamic range is higher than the typical laser output power for POF links (-3 dBm). For a longer fiber length, its higher attenuation leads to a less demanding dynamic range, as the maximum input signal level is lower for the same laser output power.

Lasers for transmission over POF are preferred class 1, that is, whose output power is limited below -4 dBm leading to eye-safety operation. Thus, installation and maintenance can be even deployed by a do-it-yourself installer, resulting in a further cost reduction, key for a widespread use in emerging applications, such as, in-car and

Fig. 5.28 Required laser
output power, estimated
by measured sensitivity and
nominal penalty sources.
Dashed line indicates the
eye-safety power level

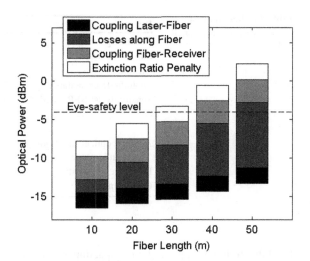

in-home networks. From the measured sensitivity, fiber losses (0.17 dB/m), and
nominal penalty values caused by fiber coupling (2 dB), receiver coupling (3 dB)
and extinction ratio (2 dB), the required laser output power of the laser can be
estimated, as shown in Fig. 5.28. Fiber length increase the required laser output
power due to higher fiber losses and sensitivity degradation caused by adaptive
equalization, which must boost the signal more to compensate the length-dependent
fiber bandwidth, and consequently, the noise is also boosted. In conclusion, eye-
safety operation is attained up to almost 30 m, while an improvement of the sensitivity is mandatory to be able to select Class 1 laser for 50 m reach.

The obtained results demonstrate the potential of POF to target multi-gigabit
transmission for short-reach applications, as error-free transmission for 1.25 Gb/s
is targeted over a single POF up to 50 m and eye-safety operation up to almost
30 m. Finally, the main performances of the POF receiver are summarized in
Table 5.6.

5.5 Conclusions

In this chapter, an optical receiver integrated in 0.18 μm CMOS technology is
presented. It is designed for low-cost applications, thus a plastic optical fiber and a
large area photodiode are used. Adaptive amplification and equalization are
mandatory to achieve high performance because of the large capacitance of the
photodiode and the frequency response of the plastic optical fiber, which strongly
depends on its length.

The receiver is based on the building blocks detailed in Chaps. 3 and 4, the
transimpedance amplifier and the post-amplifier, respectively. To achieve gigabit
speed, it was only mandatory to reduce the shunt-feedback resistance of TIA to

Table 5.6 Summary of receiver performance for SI-POF

Parameter	Value
Technology	0.18 μm CMOS
Supply voltage	1.8 V
Bit rate	1.25 Gb/s
Photodiode capacitance	$C_{PD} = 3$ pF
Sensitivity @ BER $= 10^{-12}$	−16.4 dBm @ 10 m
	−15.3 dBm @ 30 m
	−13.2 dBm @ 50 m
Average optical input power dynamic range	20 dB @ 10 m (−16.4 dBm to 3.5 dBm)
DC power dissipation	108 mW
Core chip area	0.09 mm^2

1 kΩ due to the high photodiode capacitance at the cost of sensitivity degradation. In addition, a photodiode monitor was included to control the transimpedance gain, attaining a wide input dynamic range. No modifications were necessary for PA, as it fulfilled the required noise, gain, and speed specifications. An adaptive equalizer was added between the two main blocks to compensate the frequency response of the fiber depending on its length, and to convert the single-ended signal from the TIA to a fully differential signal required by the PA.

The adaptive equalizer implementation is formed by a voltage-controlled equalizer and a servo loop, which generates the control voltage depending on the power spectrum of the data stream. The simplest solution consists of comparing the entire signal power with a low-pass filtered signal power; then the servo loop is formed by an LPF and a power error detector. The LPF is designed to provide the same signal power as the entire signal by a proper combination of gain and bandwidth. The power error detector is based on the differential pair as the output current is proportional to the squared input voltage and to the signal power.

The prototype is aimed for multi-gigabit short-range applications, targeting up to 2 Gb/s. Experimental measurements show an error-free sensitivity below −16 dBm for 1.25 Gb/s through 10 m of SI-POF, demonstrating the capability of this low-cost transmission channel. Thanks to the adaptive equalizer, the receiver is operative up to 50 m of SI-POF; however, the sensitivity is degraded to below −15 and −13 dBm for 30 and 50 m, respectively. Consequently, the laser must emit at least −1 dBm to achieve more than 3 dB margin above sensitivity level, assuming a 1 dB loss for both fiber couplings and a 7 dB loss along a fiber for the longest distance considered (50 m) for SI-POF. The power consumption of the whole receiver remains below 110 mW. Currently, new prototypes are being developed which might increase speed up to 3.125 Gb/s, targeting 10 Gb/s for Ethernet 10GBase-LX4 standard (IEEE Std 2003), improve the adaptability of the equalizer to compensate additional variations, and/or are suitable for low-voltage operation.

In the literature, there are a few examples of high speed transmission over POF, although the comparison among them is not easy. In (Zerna et al. 2009), an adaptive equalizer is presented targeting 1 Gb/s over 50 m POF channel. However, the power consumption is quite high (165 mW only for the equalizer) and the received power level for low BER is not reported. A data rate as high as 3 Gb/s over POF is reported in (Dong and Martin 2010). This speed is achieved thanks to several factors; the prototype is integrated in a nanometer CMOS technology (65 nm), a graded-index POF is employed and the longest distance considered is 30 m. Nevertheless, a strong effort is currently being made to develop POF-compliant receivers targeting gigabit data rate.

In future works, the cost can be reduced further if the photodiode is integrated; however, the drawback worsens. In addition to the large capacitance, the responsivity is much lower compared with an external photodiode and is further degraded by a slope of − 4 dB/dec due to slowly diffusing carriers (Radovanovic et al. 2003). All these frequency limitations must be compensated either by the designed equalizer, by optimized layout techniques for photodiode design or by both (Chen et al. 2007). Tavernier and Steyaert (Tavernier and Steyaert 2010) reports an 800 Mb/s speed over SI-POF for a prototype integrated in 0.18 µm CMOS technology including an integrated photodiode. This result is encouraging in order to achieve gigabit speed with a CMOS fully integrated receiver; however, the length of the POF channel is not mentioned in the paper.

References

Aznar F, Celma S, Calvo B (2010) A 0.18 µm CMOS 1.25 Gbps front-end receiver for low-cost short reach optical communications. In: Proceedings of the 36th European solid-state circuits conference, pp 554–557

Aznar F, Gaberl W, Zimmermann H (2011a) A 0.18 µm CMOS transimpedance amplifier with 26 dB dynamic range at 2.5 Gb/s. Microelectron J 42:1136–1142

Aznar F, Celma S, Calvo B (2011b) A 0.18 µm CMOS linear-in-dB AGC post-amplifier for optical communications. Microelectron Reliab 51:959–964

Budin J (1989) Determination of the zero-dispersion shift in single-mode fibers with perturbed index-profile. In: Proceedings of mediterranean electrotechnical conference, pp 583–585

Chen WZ, Huang SH, Wu GW, Liu CC, Huang YT, Chin CF, Chang WH, Juang YZ (2007) A 3.125 Gbps CMOS fully integrated optical receiver with adaptative analog equalizer. In: Proceedings of the 2007 IEEE Asian solid-state circuits conference, pp 396–399

Cheng K-H, Tsai Y-C (2010) A 5 Gb/s inductorless CMOS adaptive equalizer for PCI express generation II applications. IEEE Trans Circuits Syst II Express Briefs 57(5):324–328

Couch LW (2007) Digital and analog communication systems. Prentice Hall, Upper Saddle River

Dong Y, Martin K (2010) Analog front-end for a 3 Gb/s POF receiver. In: Proceedigns of the 2010 IEEE international symposium on circuits and systems, pp 197–200

Hall S, Heck H (2009) Advanced signal integrity for high-speed digital designs. Wiley-IEEE Press, Hoboken

Hamamatsu Photonics, Si PIN Photodiode, S5971, S5972, S5973 Series, Solid State Division, http://jp.hamamatsu.com/resources/products/ssd/pdf/s5971_etc_kpin1025e06.pdf

Hermans C, Steyaert M (2007) Broadband opto-electrical receivers in standard CMOS. Analog circuits and signal processing, Springer

IEEE Std. 802.3af-2003

Kawai S (2005) Handbook of optical interconnects. CRC Press, Boca Raton

Koonen AMJ et al (2011) A look into the future of in-building networks: radmapping the fiber invasion. In: Proceedings of the 20th international conference on plastic optical fibers, pp 41–46

Kuo K-C, Leuciuc A (2001) A linear MOS transconductor using source degeneration and adaptive biasing. IEEE Trans Circuit Syst II: Analog Digital Signal Process 48(10):937–943

Lin C-W, Liu Y-Z, Hsu KYJ (2004) A low distortion and fast settling time automatic gain control amplifier in CMOS technology. In: Proceedigns of the 2004 IEEE international symposium on circuits and systems, vol 1, pp 541–544

Liu J, Lin X (2004) Equalization in high-speed communications systems. IEEE Circuits Syst Mag 4(2):4–17

Muller P, Leblebici Y (2007) CMOS multichannel single-chip receivers for multi-gigabit optical data communications. Analog circuits and signal processing. Springer, Berlin

Nyquist H (1928) Certain topics in telegraph transmission theory. Trans Am Inst Electr Eng 47(2):617–644

Otín A (2006) A design strategy for vhf filters with digital programability. PhD thesis, University of Zaragoza, Spain

Park S-J, Lee C-H, Jeong K-T, Park H-J, Ahn J-G, Song K-H (2004) Fiber-to-the-home services based on wavelength-division-multiplexing passive optical network. J Lightwave Technol 22(11):2582–2591

Polishuk P (2006) Plastic optical fibers branch out. IEEE Commun Mag 44(9):140–148

Radovanovic S, Annema AJ, Nauta B (2003) Physical and electrical bandwidths of integrated photodiodes in standard CMOS technology. In: IEEE conference on electron devices and solid-state circuits, pp 95–98

Razavi B (2008) Fundamentals of microelectronics. Wiley, Hoboken

Säckinger E (2005) Broadband circuits for optical fiber communication. Wiley, Hoboken

Schrader JHR, Klumperink EAM, Visschers JL, Nauta B (2005) CMOS transmitter using pulse-width modulation pre-emphasis achieving 33 db loss compensation at 5 Gb/s. Digest of technical papers of 2005 symposium on VLSI circuits, pp 388–391

Shannon CE (1949) Communication in the presence of noise. Proc IRE 37(1):10–21

Sun R, Park J, O'Mahony F, Yue CP (2006) A tunable passive filter for low-power high-speed equalizers. Symposium on VLSI circuits

Sundermeyer J, Zerna C, Tan J (2009) Integrated analogue adaptive equalizer for gigabit transmission over standard step index plastic optical fibre (SI-POF). In: Proceedings of IEEE LEOS annual meeting conference, pp 195–196

Tavernier F, Steyaert M (2008) Power efficient 4.5 Gbit/s optical receiver in 130 nm CMOS with integrated photodiode. In: Proceedings of the 34th European solid-state circuits conference, pp 162–165

Tavernier F, Steyaert M (2010) A high-speed POF receiver with 1 mm integrated photodiode in 180 nm CMOS. 36th European conference and exhibition on optical communication, pp 1–3

Thorlabs Inc http://www.thorlabs.com

Zerna C, Sundermeyer J, Tan J, Fiederer A, Verwaal N (2009) Adaptive integrated equalizing techniques for SI-POF home networking links. 18th international conference on plastic optical fibers

Zhou Y, Wah MCY (2006) A wide band CMOS RF power detector. In: Proceedigns of the 2006 IEEE international symposium on circuits and systems, pp 4228–4231

Ziemann O, Krauser J, Zamzow P, Daum W (2008) POF handbook: optical short range transmission systems. Springer, Berlin

Chapter 6
Conclusions

Throughout this book, the most relevant results and main conclusions have been summarized in the final discussion of each chapter. In this final chapter, the most significant contributions will be reported to give a general overview of the entire work.

First, the main objectives presented in the first chapter will be examined, corroborating their fulfillment and leading to the corresponding conclusions. Then, further research directions will be pointed out, comprising the questions not considered in this book or a development of the ones already completed. These proposed topics could be studied in future investigations to extend the work presented here.

6.1 General Conclusions

Optical communications marked a before and after for long-haul transmissions. The electromagnetic window provided by optical fibers, whose losses are several orders of magnitude lower than that provided by copper cables, is suitable for a transmission speed over long distances only attainable by this technology. Sophisticated optimizing techniques lead to exceed, in 2009, the figure of merit of 100 Pb/s × km for the bit rate–distance product over 155 channels.

The technological evolution achieved for long-distance transmissions is being reflected in short-haul applications. Despite the inherent advantages to optical transmission, such as immunity to electromagnetic interference and galvanic separation, short-haul systems have to compete in terms of cost with electric counterparts. Thus, for applications, such as fiber-to-the-home, targeting transmission speeds higher than 100 Mb/s, or communications inside cars, the transition has begun. In future, communication buses between components in an electronic board, or even between devices in the same chip, might demand optical transmission.

F. Aznar et al., *CMOS Receiver Front-ends for Gigabit Short-Range*
Optical Communications, Analog Circuits and Signal Processing,
DOI: 10.1007/978-1-4614-3464-1_6, © Springer Science+Business Media New York 2013

A low-cost combination of the transmission channel and electronic devices is the critical condition to the viability of the system. Consequently, the plastic optical fiber (POF) and the CMOS microelectronic technology become the favorite choice. In addition to the lower fabrication cost, the POF entails an easier installation, due to its higher core diameter and is suitable for light emitters and photodetectors integrated in submicron CMOS technology. Such a technology offers a reduced fabrication cost thanks to its massive use in digital electronics, while its highest operation frequency is suitable for multi-gigabit transmissions.

Under this premise, scientific community has made a strong effort to design optical transmission systems, supported by the constant development of the microelectronic foundries and optical fibers. This book focuses on the design of an optical receiver integrated in CMOS technology with an external photodiode. The essential characteristics to optimize are the speed, determined by the transmitted data rate, the sensitivity, related to the electronic noise from the own receiver, and the input dynamic range, derived from the highest-lowest amplitude ratio of signal properly processed.

The first building block of the designed microelectronic device is the transimpedance amplifier, responsible for converting the current provided by the photodiode into a voltage. The shunt-feedback structure shows the best noise–speed trade-off. The obtained results for a low-cost 0.18 µm CMOS technology are a bit rate of 2.5 Gb/s and a sensitivity of −26 dBm, assuming a photodiode with high responsivity (1 A/W). A novel technique to extend the input dynamic range is proposed, targeting a proper operation up to 0 dBm, increasing the input dynamic range in 16 dB. The power consumption of this stage is lower than 10.6 mW. The migration to a superior 90 nm CMOS technology leads to a considerable improvement of the sensitivity up to −30 dBm and the power consumption below 4.3 mW for the same transmission speed.

The output signal of the transimpedance amplifier must be amplified to correctly discriminate between digital logical levels. This is the assignment of the second building block: the post-amplifier. Integrated in a low-cost 0.18 µm CMOS technology, the design offers low input-referred noise, a bandwidth at least 1.5 times higher than that from the transimpedance amplifier with a programmable gain from 33 to −3 dB with a 6 dB step. Therefore, the signal fulfills the requirements demanded by the clock and data recovery circuit while the post-amplifier does not excessively degrade the signal in terms of sensitivity or speed. The final prototype includes an automatic gain control with a worst-case settling time lower than 1 µs and an offset compensation loop with a low-frequency cut-off of 100 kHz, while the power consumption remains below 58 mW. Experimental measurements corroborate the simulation results with a maximum error for the gain of 1 dB and a bandwidth shrinkage of 15 %, what would degrade slightly the output signal of the entire receiver.

Both the aforementioned blocks have been integrated together, comprising an optical receiver for step-index plastic optical fiber with 1 mm core. Such a receiver is restricted to short-haul applications due to the losses of the plastic optical fiber (0.14 dB/m) and, specially, to its bandwidth-length dependency (40 MHz × 100 m).

Consequently, including a continuous-time equalizer to the receiver chain is mandatory to compensate the limited fiber bandwidth, and then targeting gigabit speed. In addition, due to the diameter core of the fiber, a relatively large photodiode is necessary to efficiently detect the light, thus the receiver must be able to handle a high capacitance at the input. The prototype targets a speed of 1.25 Gb/s for POF with a length of 50 m, and a sensitivity below −13 dBm with a commercial external photodiode.

As a final conclusion, it must be noted the potential of the systems based on POF and submicron CMOS technology to provide the final user with a considerable speed increase compared to electric systems, in addition to the complete confidentiality of the information thanks to the immunity to electromagnetic interference.

6.2 Further Research Directions

Recent research, including this work, has clearly demonstrated that CMOS has matured to a technology that is capable of competing with bipolar or even GaAs technologies in the area of optical communication up to gigabit data rates. However, there are several challenges, which have not been treated here, but will most certainly be interesting to study in future projects.

The proposed transimpedance amplifier and the post-amplifier core are able to operate at lower supply voltages than typical for 0.18 μm CMOS due to only two transistors being cascaded from supply to ground. It was demonstrated that the operation of the 90 nm TIA at 1 V targets better sensitivity. The reduction of supply voltage for 0.18 μm CMOS leads to reduction in power dissipation. In order to integrate a low-voltage optical receiver for low-cost applications, where the frequency limitation due to a large photodiode or the optical fiber must be compensated, new architectures for equalizers suitable for low-voltage operation are mandatory.

Another interesting research direction faces with the integration of photodiodes in the same substrate as the front-end. In addition to the benefits in cost of a complete integrated receiver, it must be remarked the increase of reliability and the optimization of the connection between the photodiode and the front-end thanks to avoid bound wires. In contrast, integrated photodiodes for visible range show worse performance than external ones, what is partially compensated by a careful design and/or a more complex front-end.

The full integration of the analog front-end with digital circuitry may entail a non-considered issue. The noise caused by a huge number of transitions during digital processing could affect the sensitivity of the receiver. Although a basic isolation was employed for the core of the transimpedance amplifier, which is the most sensitive to noise, more research on isolation techniques could be necessary.

The performances of the integrated prototype could vary from the nominal value due to several factors (fabrication process, mismatching, aging, supply voltage...).

The minimization of these variations or at least the optimization of the worst case is an interesting research field, as the receiver should properly operate under a wide range of conditions. Especially important is the impact of the temperature on the receiver performances, as the power dissipated by the receiver itself or the surrounding circuits could heat the substrate up to 80 °C or even more.

Appendix

Measurement Considerations

To verify simulation results several circuits were measured. This section introduces the theory and techniques that were used to realize the measurements. First of all, here is a layout of the equipment employed, apart from the typical measuring equipment found in any laboratory:

- Network analyzer: Rohde & Schwartz ZVL6 (9 kHz–6 GHz).
- Digital communications analyzer: Agilent 86100C.
- Bit error ratio tester: Agilent N4906A.

S Parameters

High frequency circuits typically are characterized by scattering (S) parameters (Säckinger 2005). The reason for preferring them is that they are easier to measure, as 50 Ω terminations are required instead of open or short conditions, possibly leading to instability, and they are based on incident and outgoing traveling waves, easily and accurately measured with directional couplers. In practice, a so-called network analyzer is used to measure S parameters.

The key idea behind the S parameters is to decompose each port voltage (v_i) and port current (i_i) into an incident and outgoing component that correspond to the incident and outgoing waves on a transmission line, as illustrated in Fig. A.1.

The decomposition is done according to three conditions: the superposition of the incident and outgoing voltage component equals the port voltage, the superposition of the incident and outgoing current component equals the port current, and voltage-to-current ratio of each component is equal to the characteristic impedance of the transmission line, R_0. A purely resistive

F. Aznar et al., *CMOS Receiver Front-ends for Gigabit Short-Range*
Optical Communications, Analog Circuits and Signal Processing,
DOI: 10.1007/978-1-4614-3464-1, © Springer Science+Business Media New York 2013

Fig. A.1 Input and output
signals seen as incident and
outgoing waves

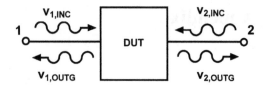

characteristic impedance of 50 Ω is assumed for the rest of this discussion. This
can be written as

$$v_i = v_{i,INC} + v_{i,OUTG} \tag{A.1}$$

$$i_i = i_{i,INC} - i_{i,OUTG} \tag{A.2}$$

$$R_0 = \frac{v_{i,INC}}{i_{i,INC}} = \frac{v_{i,OUTG}}{i_{i,OUTG}} \tag{A.3}$$

where the sub index i refers to the port. For a two port system, where port 1 is the
input port and port 2 is the output port by convention, the S parameters are defined
as the following ratios:

$$S_{11} = \frac{v_{1,OUTG}}{v_{1,INC}} = \frac{v_1 - R_0 i_1}{v_1 + R_0 i_1} \tag{A.4}$$

$$S_{21} = \frac{v_{2,OUTG}}{v_{1,INC}} = \frac{v_2 - R_0 i_2}{v_1 + R_0 i_1} \tag{A.5}$$

$$S_{12} = \frac{v_{1,OUTG}}{v_{2,INC}} = \frac{v_1 - R_0 i_1}{v_2 + R_0 i_2} \tag{A.6}$$

$$S_{22} = \frac{v_{2,OUTG}}{v_{2,INC}} = \frac{v_2 - R_0 i_2}{v_2 + R_0 i_2} \tag{A.7}$$

Thus, $S_{\mu v}$ parameter is the ratio between the outgoing wave measured in port μ
and the incident wave exciting port v. Note that the outgoing wave can be a
reflected or a transmitted wave. In addition, this definition of S parameters is based
on voltage waves, but a definition based on incident and outgoing current
components would yield to identical S parameters. In fact, S parameters usually
are defined in the literature in terms of the so-called power waves, proportional to
the defined voltage waves. Previous equations can be grouped using a matrix:

$$\begin{pmatrix} v_{2,INC} \\ v_{2,OUTG} \end{pmatrix} = \begin{pmatrix} S_{11} & S_{12} \\ S_{21} & S_{22} \end{pmatrix} \begin{pmatrix} v_{1,INC} \\ v_{1,OUTG} \end{pmatrix} \tag{A.8}$$

With this notation, a generalization to an N-port system can be easily written. In
our case, in spite of being a 4-port system, it is limited to 2-ports thanks to only
consider differential signals.

The S_{11} parameter is known as the input reflection coefficient because it
describes what fraction of an incident wave traveling on an ideal 50 Ω

Fig. A.2 Calibration kit Rohde & Schwartz ZV-Z132

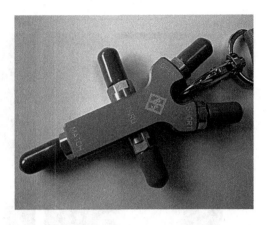

transmission line is reflected back from the input port. Equivalently, the S_{22} parameter is known as the output reflection coefficient. Therefore, both parameters are a measure of how close the input and output impedances are to the ideal 50 Ω value. In fact, the input Z_{IN} and output Z_{OUT} impedances can be derived from the measured S_{11} and S_{22} respectively:

$$Z_{IN} = R_0 \frac{1 + S_{11}}{1 - S_{11}} \tag{A.9}$$

$$Z_{OUT} = R_0 \frac{1 + S_{22}}{1 - S_{22}} \tag{A.10}$$

The S_{21} parameter is knows as the forward transmission coefficient and is closely related to the loaded voltage gain A_V. However, unlike the loaded voltage gain, S_{21} also depends on the quality of the input matching. The next relationship can be obtained:

$$A_V = \frac{S_{21}}{1 + S_{11}} \tag{A.11}$$

Finally, the S_{12} parameter is known as the reverse transmission coefficient and for amplifier designs, its value is only a measure of how close is the prototype to a unilateral amplifier, that is, an amplifier with no reverse transmission.

Calibration

In general, calibration is a measurement of a known magnitude, denominated standard. Thus, a possible deviation of the measurements, due to environmental factors of aging process for instance, can be detected and, usually, corrected. Then, the device under test (DUT) can be measured accurately compared with the standard.

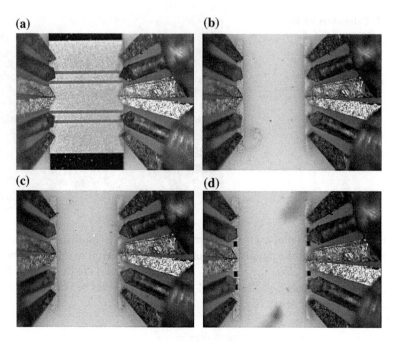

Fig. A.3 Calibration process: **a** through, **b** short, **c** open, **d** load

An international standard is mandatory to compare measurement properly. As a curiosity, kilogram is the only SI (*Système international d'unités*) unit that is still defined by an artifact, whereas all other SI units have been redefined using a fundamental physical property that can be reproduced in adequately equipped laboratories. The kilogram is currently defined as being equal to the mass of the international prototype kilogram, an alloy of iridium and platinum which is saved in the *Bureau International des Poids et Mesures* (BIPM), in Sèvres, near Paris, France. A redefinition based on the Plank's constant or the Avogadro number is under study.

For frequency-domain characterization, a complete calibration process for a 2-port system consists of four measurements: short, open, load and through (SOLT). Network analyzer is calibrated with a SOLT kit. An example of SOLT kit is illustrated in Fig. A.2.

After calibration, the reference plane is defined to measure the DUT. For on-wafer technique, the accuracy goes one step beyond. A substrate provides the SOLT references, so the calibration is achieved from the probes, not only the network analyzer. Figure A.3 illustrates the four steps to calibrate differential GSGSG probes with a calibration substrate.

Figure A.4 shows a typical result of SOLT measurements. The frequency range and the exciting signal power must be chosen before calibration.

As seen, the system is basically symmetric as S_{11}–S_{22} and S_{12}–S_{21} pairs show a really similar graph. S_{11} and S_{22} parameters are represented in Smith charts while

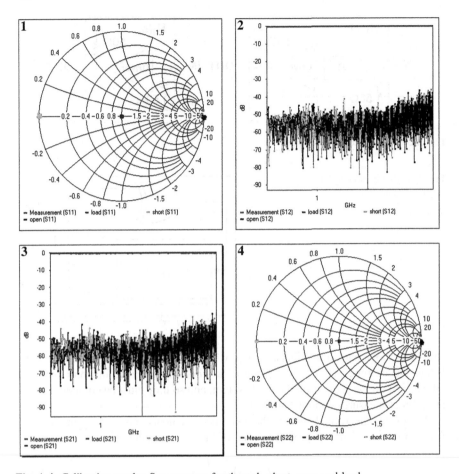

Fig. A.4 Calibration results: S parameters for through, short, open and load

S_{12} and S_{21} are represented in magnitude Bode plots with dB-scale. Calibration results show a measurement range from reference near 40 dB and a good matching between equipment and calibration subtract for the selected frequency range.

De-Embedding

In addition to the calibration, another technique must be implemented to measure the real frequency performance of the device itself. If parasitic effects of the rest of necessary components degrade significantly the frequency response of DUT, compensation can be achieved thanks to de-embedding technique.

50 Ω ports require the implementation of matched resistors and an output driver to provide sufficient output swing. These mandatory components and the PAD

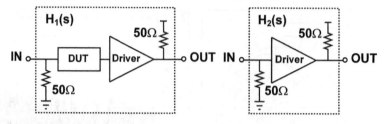

Fig. A.5 De-embedding technique

Fig. A.6 Pseudorandom bit
sequence generator

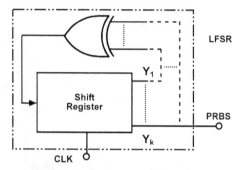

structure show parasitic effects, which will be included in the measurements. When the designed device is a piece of a whole system, the inherent parasitic effects will not be present on the final system, and so they must be compensated by an indirect measure without the DUT.

Therefore, the ratio between the two measurements represented in Fig. A.5 leads to the real frequency performance of the device under test.

$$H_{DUT}(s) = \frac{H_1(s)}{H_2(s)} \qquad (A.12)$$

PRBS Generator

A typical PRBS generator is formed by a linear feedback shift register (LFSR) (Alfke 1996) including a multiple-input XOR gate, as shown in Fig. A.6. The sequence is generated at the output of the register while its input is fed back with a combination for the inputs of the XOR gate.

The operation principle of such a sequential circuit can be represented by a state diagram, as shown Fig. A.7. It illustrates how the digital word $Y = \{Y_1 Y_2 ... Y_k\}$ changes when CLK indicates. If the initial state is reached again after N states, the generator provides a repetitive pattern of N bits, where $a_j \equiv Y_k$, as the last bit of

Fig. A.7 State diagram of a
PRBS generator

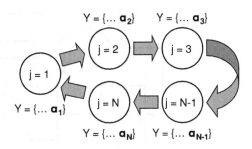

Table A.1 Maximal length sequence properties

k	XOR from	Pattern length: N		N_{MAX}	Number of runs: r		Disparity (%)
3	Y_3, Y_2	2^3-1	7	3	2^2	4	14.29
7	Y_7, Y_6	2^7-1	127	7	2^6	64	0.79
15	Y_{15}, Y_{14}	$2^{15}-1$	32,767	15	2^{14}	16,384	3.05E-03
23	Y_{23}, Y_{18}	$2^{23}-1$	8,388,607	23	2^{22}	4,194,304	1.19E-05
31	Y_{31}, Y_{28}	$2^{31}-1$	2,147,483,647	31	2^{30}	1,073,741,824	4.66E-08

the digital word is usually chosen as output bit.

The state formed by all zeros ('000....000') is not included in the sequence because is stable itself. That is why the length of the longest sequence is limited to

$$N = 2^k - 1 \qquad (A.13)$$

where k is the number of outputs of the shift register, which match with the number of flip-flops forming it. It can be demonstrated that, independent on the number of bistables, the sequence generated for a particular feedback is a PRBS signal and reaches maximal length sequence (MLS). They are usually denominated by the number of flip-flops, for example, PRBS31. In Table A.1, some properties of MLS for typical number of flip-flops are presented.

Notice that for all this cases, only a 2-input XOR gate is required, while for a general case a 4-input XOR gate may be necessary. N_{MAX}, which match with the number of bistables k, is the maximum number of bits with the same state, that is, the longest run, and the number of runs within the pattern can be calculated from

$$r = 2^{k-1} \qquad (A.14)$$

Furthermore, a regular distribution of the length of the runs is expected. If r is the number of runs within the pattern, $r/2$ runs has length 1, $r/4$ has length 2, etc., up to an only run with N_{MAX}, always formed by states '1'. Finally, the repetitive pattern is formed by one '1' more than '0', because the only state not present in the sequence consists of all zeros ('000...000'), so the generated PRBS is a particular case with

$$c = \frac{1}{2} \Rightarrow D = \frac{1}{N} \qquad (A.15)$$

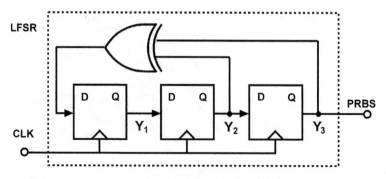

Fig. A.8 Pseudorandom bit sequence generator with $k = 3$

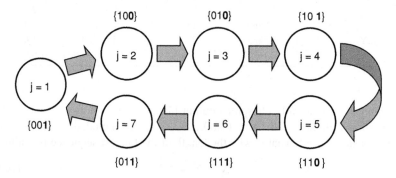

Fig. A.9 State diagram of a PRBS generator with $k = 3$

Table A.2 States for a LFSR with $k = 3$

j	Y_1	Y_2	$Y_3 = a_j$
1	0	0	1
2	1	0	0
3	0	1	0
4	1	0	1
5	1	1	0
6	1	1	1
7	0	1	1

minimizing the disparity caused by an odd number of bits. To conclude this introduction to PRBS signals, some simple examples will be shown. For $k = 3$, the PRBS generator and its corresponding state diagram are shown in Figs. A.8 and A.9 respectively.

As shown in Table A.2, the generated PRBS sequence ('1001011') confirms the results provided in Table A.1 and the no correlation among the bit of the sequence, calculated by (2.11). Due to the simplicity of the generated pattern with $k = 3$, 7 is selected to demonstrate the run distribution. With an equivalent initial digital word $Y = \{0000001\}$, the generated sequence is ('1000000100000110000101000011111

Table A.3 Run distribution for $k = 7$

Run length	Number of runs	Run state
1	32	16 "1" and 16 "0"
2	16	8 "1" and 8 "0"
3	8	4 "1" and 4 "0"
4	4	2 "1" and 2 "0"
5	2	"1" and "0"
6	1	Always "0"
$7 = N_{MAX}$	1	Always "1"

Table A.4 Comparison between chosen CMOS technologies

Parameter	Units	UMC	
Process name	–	L180 MM/RF	L90SP
Substrate type	–	P-substrate	
Number of poly/metal layers	#	1P6M	1P9M
Core devices	V	1.8	1
Operating voltage	μm	0.18	0.08
Min gate length	V	0.51/−0.5	0.33/−0.277
V_{TH} N/P			
Available IO devices	V	3.3	2.5
Operating voltage	μm	0.34	0.24
Min gate length	V	0.65/−0.7	0.548/−0.5
V_{TH} N/P			
V_{TH} options	–	LVT	LVT/HVT
High ohmic resistor (HR)	Ω/sq	1,039	1,012
Metal metal cap (MiM cap)	fF/μm	1	1.544

001000 10110011101010011111010000111000100100110110101101111011000
11010010111 011100110010101010111111'), what leads to a run distribution as
shown in Table A.3.

So the run distribution consists of two long runs, one formed by '1' with length
N_{MAX} and another formed by '0' with length $N_{MAX}-1$, and a combination of runs
formed by '1' or '0' with every possible length.

Technological Parameters

This section summarizes the most important parameters associated to CMOS
technologies, 0.18 μm and 90 nm, considered in this work both from UMC
(United Microelectronics Corporation).[1] First, Table A.4 compares both

---

[1] http://www.umc.com/English/process/index.asp

Table A.5 Key device parameters

Device type	Core		I/O	
Parameter		LVT		LVT
V_{CC} (V)	1.8		3.3	
T_{OX} (A)	33		65	
Lmin_draw (μm)	0.18		0.34	
V_{TH} N/P(V)	0.51/−0.5	0.22/−0.22	0.65/−0.7	0.31/−0.42
I_{DS} N/P (μA/μm)	625/−244.2	720/−270	590/−260	640/−250
I_{OFF} N/P (A/μm)	7.6p/−8.1p	29.4n/−12.4n	1p/−0.5p	0.92n/−2p
Gate delay (ps/stage)	28.5	–	55	–

Table A.6 Key design rules

Layers	Min. width (μm)	Min. spacing (μm)	Pitch (μm)
Diffusion	0.24	0.28	0.52
Inter-well	0.9	0.9	1.8
Drawn poly	0.18	0.28	0.46
Contact	0.24	0.26	0.5
Metal 1	0.24	0.24	0.48
MVia1–MVia5	0.28	0.28	0.56
Inter metal	0.28	0.28	0.56
Metal_cap	0.6	0.55	1.15
Metal 6 8 K	0.44	0.44	0.88
Metal 6 12 K	0.8	0.8	1.6
Metal 6 20 K	1.2	1.0	2.2

technologies, whereas Sects. A.3.1 and A.3.2 report a summary of each one.

UMC 0.18 μm Mixed-Mode/RF CMOS Process

UMC L180 MM/RF 1.8 V/3.3 V 1P6 M technology is the CMOS process based on general P-Sub structure with 1 layer of poly, 6 layers of aluminum metal and FSG dielectrics. Moreover, the MM/RF process also includes several optional layers which defined and decided by customer's applications design (Tables A.5, A.6).

Key Process Features

- P-Sub CMOS process with optional Deep N-well.
- Dual gate oxide thickness (1.8 V/3.3 V).
- Mixed optional device application is available.
- Three M6 aluminum thickness types depend on customer's design application.
- FSG dielectrics.

Table A.7 Key core devices parameters

Device type	SP			LL		
Parameter	RVT	HVT	LVT	RVT	HVT	LVT
V_{CC} (V)		1			1.2	
T_{OX} (A)		16			22	
Lmin_draw (μm)		0.08			0.09	
V_{TH} N/P(V)	0.24/−0.177	0.37/−0.31	0.155/−0.105	0.45/−0.406	0.54/−0.455	0.381/−0.284
I_{ON} N/P (μA/μm)	655/−280	480/−195	760/−320	480/−185	385/−155	552/−250
I_{OFF} N/P (A/μm)	5n/−10n	0.3n/−0.4n	80n/−80n	30p/−30p	6p/−10p	0.2n/−0.8n
Gate delay without parasitic RC loading (ps/stage)	10.6	16.1	8.6	17.5	21.3	12.8
Over drive feasibility (V)		1.2		−	−	−

Table A.8 Key I/O devices parameters

Device type	I/O		
Parameter	1.8 V	2.5 V	3.3 V
V_{CC} (V)	1.8	2.5	3.3
T_{OX} (A)	31	52	65
Lmin_draw (μm) N/P	0.18/0.18	0.24/0.24	0.34/0.34
V_{TH} N/P(V)	0.465/−0.365	0.44/−0.42	0.506/−0.49
I_{ON} N/P (μA/μm)	600/−255	600/−275	580/−260
I_{OFF} N/P (A/μm)	10/−25	15/−15	10/−10
Gate delay without parasitic RC loading (ps/stage)	22	24.7	39.4

UMC 90 nm Logic and Mixed-Mode Process

UMC L90 1P9 M Low-k platform is a 90 nm generation CMOS process technology based on P-Sub wafer, twin well structure with triple well option, 9 layers of copper interconnects, and Low-k dielectrics.

L90 platform includes 2 families of core devices, SP (Standard Performance) and LL (Low Leakage), which can stand alone by itself or integrated onto a single chip. Each family includes 3 pair of devices, HVT, RVT and LVT, consequently, any combinations of the six pairs are allowed in a given design. In addition, three choices of thick oxide (I/O) devices are available, 1.8, 2.5 and 3.3 V, which only one can be chosen. UMC's platform also offers MIMCAP (Metal Insulator Metal Capacitor) and 32.5 KA top metal options for Mixed Signal or RF applications (Tables A.7, A.8, A.9).

Table A.9 Key design rules

Layers	Min. width (μm)	Min. spacing (μm)	Pitch (μm)
Diffusion	0.12	0.14	0.26
Inter-well: N +/NW or P +/PW	–	0.21	0.42
Drawn poly	0.08	0.16	0.24
Contact	0.12	0.16	0.28
Metal 1	0.12	0.12	0.24
1X via (via1–via5)	0.14	0.18	0.32
1X metal (metal2–metal6)	0.14	0.14	0.28
2X via (via6–via7)	0.28	0.28	0.56
2X metal (metal7–metal8)	0.28	0.28	0.56
4X via (via8)	0.56	0.56	1.12
4X metal (metal9)	0.56	0.56	1.12

Key Process Features

- P-Sub wafer, Twin Well CMOS process with Triple Well options.
- 193 nm lithography (4 layers adopted: Diff, Poly, Cont, M1).
- Six pairs of core V_{TH} and single thick oxide (I/O) devices with multiple I/O voltage options.
- Dual Poly Gate with CoSi2 process.
- Up to 1P9 M Cu: Low-k for 1X, and FSG for 2X and 4X metal layers.
- Wire Bond/Flip Chip.
- Split Word Line 1.16 and 0.99 μm2 dense SRAM Bit Cell.

References

Alfke P (1996) Efficient shift registers, LFSR counters, and long pseudo-random sequence generators, Xilinx Application Note 052. http://www.xilinx.com/support/documentation/application_notes/xapp052.pdf

Säckinger E (2005) Broadband circuits for optical fiber communication. Wiley, Hoboken

Index

F. Aznar et al., *CMOS Receiver Front-ends for Gigabit Short-Range*
Optical Communications, Analog Circuits and Signal Processing,
DOI: 10.1007/978-1-4614-3464-1, © Springer Science+Business Media New York 2013